T0321233

Cambridge Elements

Elements in the Philosophy of Biology
edited by
Grant Ramsey
KU Leuven
Michael Ruse
Florida State University

STRUCTURE AND FUNCTION

Rose Novick
University of Washington

Shaftesbury Road, Cambridge CB2 8EA, United Kingdom

One Liberty Plaza, 20th Floor, New York, NY 10006, USA

477 Williamstown Road, Port Melbourne, VIC 3207, Australia

314–321, 3rd Floor, Plot 3, Splendor Forum, Jasola District Centre, New Delhi – 110025, India

103 Penang Road, #05–06/07, Visioncrest Commercial, Singapore 238467

Cambridge University Press is part of Cambridge University Press & Assessment, a department of the University of Cambridge.

We share the University's mission to contribute to society through the pursuit of education, learning and research at the highest international levels of excellence.

www.cambridge.org
Information on this title: www.cambridge.org/9781009387088

DOI: 10.1017/9781009028745

First published 2023

A catalogue record for this publication is available from the British Library.

ISBN 978-1-009-38708-8 Hardback
ISBN 978-1-009-01388-8 Paperback
ISSN 2515-1126 (online)
ISSN 2515-1118 (print)

Structure and Function

Elements in the Philosophy of Biology

DOI: 10.1017/9781009028745
First published online: February 2023

Rose Novick
University of Washington

Author for correspondence: Rose Novick, amnovick@uw.edu

Abstract: The history of biology is mottled with disputes between two distinct approaches to the organic world: structuralism and functionalism. Their persistence across radical theory change makes them difficult to characterize: the characterization must be abstract enough to capture biologists with diverse theoretical commitments, yet not so abstract as to be vacuous. This Element develops a novel account of structuralism and functionalism in terms of explanatory strategies (Section 2). This reveals the possibility of integrating the two strategies; the explanatory successes of evolutionary-developmental biology essentially depend on such integration (Section 3). Neither explanatory strategy is universally subordinate to the other, though subordination with respect to particular explanatory tasks is possible (Section 4). Beyond structuralism and functionalism, philosophical analysis that centers explanatory strategies can illuminate conflicts within evolutionary theory more generally (Section 5).

Keywords: structuralism, functionalism, evolutionary-developmental biology, explanatory strategies, evolution

ISBNs: 9781009387088 (HB), 9781009013888 (PB), 9781009028745 (OC)
ISSNs: 2515-1126 (online), 2515-1118 (print)

Contents

1 Structure and Function 1

2 Explanatory Strategies 7

3 Understanding Evo-Devo 27

4 Integration without Subordination 40

5 Gentle Polemics 55

References 65

The forms of the spicules are the result of adaptation to the requirements of the sponge as a whole, produced by the action of natural selection upon variation in every direction.

Edward A.Minchin (1898, 569)

It would scarcely be possible to illustrate more briefly and more cogently than by these few words [. . .] the fundamental difference between the Darwinian conception of the causation and determination of Form, and that which is based on, and characteristic of, the physical sciences.

D'Arcy Thompson (1992 [1942], 693), commenting on Minchin

1 Structure and Function

1.1 The Cuvier-Geoffroy Dispute

In October 1829, Pierre-Stanislas Meyranx and Laurencet (the latter so obscure his full name remains unknown) submitted, to the Académie des Sciences in Paris, a memoir purporting to demonstrate a deep unity of form linking vertebrates and mollusks (following Appel 1987, chap. 6). Take a vertebrate. Bend it backward so that the nape touches the anus – the arrangement of its internal organs now matches, they argued, that of a cephalopod. After a long wait, Pierre André Latreille and Étienne Geoffroy Saint-Hilaire were assigned to prepare a report on the memoir. Delivered on February 15, 1830, their report brought to a boil long-simmering tensions between Geoffroy and his erstwhile friend Georges Cuvier.

One of their deepest disagreements concerned whether anatomy should be organized around Unity of Type (Geoffroy) or Conditions of Existence (Cuvier) – "whether animal structure ought to be explained primarily by reference to function or by morphological laws" (Appel 1987, 2). In 1812, Cuvier separated the animal kingdom into four *embranchements*, each characterized by a unique arrangement of functionally integrated parts. Individual species realized this arrangement in a manner befitting their particular form of life. Structural correspondences could be traced within *embranchements*, but not between them.

Geoffroy's "philosophical anatomy," by contrast, sought correspondences spanning the entire animal kingdom. Notoriously, he claimed that vertebrates are arthropods turned upside-down – a hypothesis whose apparent absurdity has not stopped it from receiving "a measure of molecular support" (de Robertis 2008). Meyranx and Laurencet's memoir enabled Geoffroy to connect vertebrates to mollusks as well. These correspondences pointed toward general morphological laws inexplicable by species' particular conditions of existence.

The official debate before the Académie ended in April 1830, but the disagreement persisted until Cuvier's death in May 1832. Though Cuvier was widely recognized as the "winner" of the debate, Geoffroy's views remained influential – his "loss" hardly dampened the prospects of philosophical anatomy. Indeed, conflicts between structuralist and functionalist visions of biology, between Unity of Type and Conditions of Existence, are persistent presences haunting the history of biology, from its origins in ancient Greece right down to contemporary debates within evolutionary theory. This Element aims to make sense of them.

1.2 Structuralism and Functionalism I

Historians, philosophers, and biologists alike have analyzed the history of biology (in whole or part) in terms of conflicts between structuralists and functionalists (Coleman 1977 [1971]; Russell 1982 [1912]; Asma 1996; Amundson 2005). Shortly after the Cuvier-Geoffroy debate, similar tensions arose between Richard Owen (e.g., 2007 [1849]) and Charles Darwin (e.g., 1964 [1859]). In the twentieth century, the functionalist modern synthesis (Dobzhansky 1982 [1937]; Mayr 1982 [1942]; Simpson 1984 [1944]) was challenged by the structuralist alternatives of Richard Goldschmidt (1982 [1940]) and Otto Schindewolf (1993 [1950]). The rise of evolutionary-developmental biology (evo-devo) generated and continues to generate structure/function disputes (Hughes and Lambert 1984; Alberch 1989; Amundson 2005; Wagner 2014). Related conflicts arise from biophysical approaches to organic form (Webster and Goodwin 1982, 1996; Ho and Saunders 1993; Kauffman 1993; Newman and Bhat 2008), which recapitulate themes from the Thompson/Minchin conflict quoted in the epigraph. Going backward in time, Empedocles' explanation of plant growth in terms of their earthen roots and fiery trunks offered an early structuralist view, against which Aristotle raised a functionalist critique (Irwin and Fine 1996, 87–88). The list could be extended *ad nauseam.*

My decision to rattle off examples before saying what structuralism and functionalism *are* is, perhaps, frustrating. But to say what they are is difficult, for they have survived thousands of years of drastic theoretical change. If structuralism is a meaningful category, it must capture how Empedocles, Richard Owen, D'Arcy Wentworth Thompson, and Pere Alberch (*inter alia*) are similar; if functionalism is a meaningful category, it must capture how Aristotle, Georges Cuvier, Ernst Mayr, and Eva Jablonka (*inter alia*) are similar. And these similarities must not be so vague as to render the designations vacuous.

This Element develops a novel account of structuralism and functionalism. The main difficulty, as I see it, stems from analyzing structuralism and functionalism as *positions* of some strange kind: contentful, yet able to survive even radical theoretical change. I focus instead on structuralist and functionalist *explanatory strategies* – abstract schemas that, uncommitted to any particular theory, capture underlying patterns in the explanation of organismic form.

Like most work on this topic, my analysis focuses on explanations of morphological form in multicellular organisms. However, structure/function disputes themselves are not so limited. They arise, for instance, in the study of genome evolution, for the genome "has an evolutionary life of its own" (West-Eberhard 2003, 19), as well as in microbial evolution (e.g., Sapp 2009, 294–99). They may even arise in the study of biological function itself, in conflicts between organizational (Mossio, Saborido, and Moreno 2009) and selected effect (Garson 2017) accounts thereof. Space constraints have narrowed my vision, but I hope my analysis will prove valuable beyond the study of multicellular form.

1.3 Three Approaches: Stances, Paradigms, Strategies

Philosophical analysis of structuralism and functionalism has two core aims: to *explain* features of the history of biology and to *provide guidance* to those who find themselves in present-day structure/function disputes. These can be further specified in terms of five desiderata. An analysis of structuralism and functionalism should

1. capture what all structuralists have in common, in virtue of which they are structuralists, and likewise for functionalists.
2. be sufficiently contentful to explain why structuralist and functionalist biologists behave as they do.
3. explain why structuralism and functionalism have persisted across radical theory change in biology.
4. clarify how, why, where, and what kind of empirical evidence is relevant to structure/function disputes.
5. explain, not just why structuralism and functionalism conflict, but also how they can be and have been integrated.

The first two desiderata capture basic constraints on explanatory adequacy. The first is intuitive: for any account to qualify as an account of structuralism and functionalism at all, it must tell us what structuralism and functionalism *are*. The second is comparably intuitive: for the account to be enlightening, the characterizations offered should allow us, not merely to *identify* structuralists and functionalists, but to explain why they engage in biological inquiry as they do.

The latter three desiderata identify particular explanatory targets. The third concerns the fact that structuralism and functionalism *have* persisted across radical theory change. We need an explanation of how this is possible and why it has happened. The fourth concerns the fact that biologists engaging in structure/function disputes bring empirical evidence to bear on them. Even if structure/function disputes have a nonempirical core, we still need to understand how and why empirical evidence becomes relevant.

The fifth concerns the fact that interactions between structuralism and functionalism are not exclusively antagonistic. Their integration has occupied less philosophical attention than their conflict; accordingly, I devote significant space to documenting that it occurs (Section 3). Even without such documentation, however, the long-term persistence of structuralism and functionalism should lead us to expect that both capture important aspects of the biological world and thus that it should be possible to integrate them.

Satisfying all five desiderata simultaneously is challenging. The first two pull in opposite directions. To satisfy the first, an account must be sufficiently abstract to capture thinkers separated by deep theoretical gulfs – but this abstraction must not cost the account its explanatory power. The third and fourth likewise exert opposed forces. To satisfy the third, an account must explain why empirical evidence has not put an end to structure/function disputes – but this explanation must not render empirical evidence entirely irrelevant. We may hope to find an acceptable compromise between both sets of opposing pulls, but it is a difficult tightrope to walk.

Until all five are satisfied, our understanding of structuralism and functionalism is lacking. *How* they are to be satisfied – whether by a single account or in a more piecemeal fashion – remains an open question. I will argue that, by treating structuralism and functionalism not as *positions* (or similar) held by particular inquirers, but rather as *explanatory strategies*, we can satisfy all five. Moreover, my account augments the explanatory power of two recent analyses (Boucher 2015; Winther 2015) that both capture important aspects of the issue, but that are not sufficient either individually or jointly.

Boucher (2015) analyzes structuralism and functionalism as stances (Van Fraassen 2002; Boucher 2014). Stances are clusters of attitudes, not sets of beliefs – they are not propositional and not truth apt. The adoption of stances is driven by one's values (both epistemic and not) and is justified pragmatically. On this view, structuralists and functionalists are distinguished by how they *approach* the organic world, which shapes but does not determine what they *believe* about it. Their disagreement concerns which features of the organic world are most important to capture in biological theorizing. Functionalists take explaining organism-environment fit as a central explanatory task, while structuralists are more concerned with deep similarities between species living in rather different environments.

Boucher's analysis satisfies the first and third desiderata. It identifies a specific feature – adoption of a particular stance – by which structuralists and functionalists can be identified and distinguished. Moreover, because adopting a stance does not require endorsing any particular belief, stances can survive even radical theoretical change. Thus Boucher can explain empirical advances have failed end the debate. Though not uninfluenced by empirical evidence, stances are not (dis)confirmed by it.

The second, fourth, and fifth desiderata are trickier. Stances capture what all functionalists (structuralists) have in common, but highly abstractly, which limits what Boucher can explain. Knowing merely that a biologist adopts a particular stance, one cannot predict much about their research practices. Precisely because stances have a loose, nonlogical relationship with theories and evidence, Boucher's account says little about what types of evidence functionalists (structuralists) will gather, what kinds of theories they will develop on that basis, what kinds of explanations they will favor, and how they will conduct empirical disputes. Moreover, because structuralist and functionalist stances involve incompatible values (one cannot simultaneously foreground and background adaptation), stances cannot be integrated. Accordingly, Boucher (2019) limits nonantagonistic interactions between structuralism and functionalism to pluralistic tolerance.

Winther (2015) analyzes structuralism and functionalism (or "adaptationism") as Kuhnian paradigms. Paradigms include diverse elements (see Winther 2015, 472 for a full list); most important here is that they include *both* theoretical and empirical *and* nonpropositional commitments, including explanatory standards, research questions, and methods. Proponents of the functionalist paradigm treat organism–environment fit as the central evolutionary problem and natural selection as the most important explanatory resource for solving it; they may also endorse methodological adaptationism (Godfrey-Smith 2001). Proponents of the structuralist paradigm are most concerned with problems of how structures form and emphasize the role of "mathematical laws of development and physiochemical morphogenetic mechanisms" in explaining this (Winther 2015, 473).

Compared to Boucher's, Winther's analysis has inverse virtues and vices. Because paradigms involve structured relationships between value judgments (including those characteristic of Boucher's stances), methods, and particular beliefs, Winther's analysis does an excellent job explaining why particular biologists conduct inquiry as they do, including why they seek out particular kinds of evidence. Moreover, while distinct paradigms may involve incompatible values, their other elements can be complementary, so Winther (2015, sec. 21.3.3) has the resources to explain not just "imperialist" but also "collaborative" interactions.

Winther's analysis purchases these virtues at the cost of generality. Whereas stances are persistent by design, paradigms are ephemeral by design. They characterize complex commitments of particular scientific communities over comparatively short temporal durations. The structuralist and functionalist paradigms Winther identifies are specific to contemporary evolutionary theorizing. Nor does Winther's account answer what it is that makes these paradigms structuralist and functionalist, respectively – what they have in common with past structuralist and functionalist paradigms. This is not a problem for Winther: his concern is to understand evo-devo, and paradigms are an appropriate tool. However, paradigms cannot furnish a *general* account of structuralism and functionalism.

Can we satisfy all five desiderata by conjoining the two accounts? The idea here is that scientists who adopt structuralist (functionalist) stances develop particular structuralist (functionalist) paradigms. What structuralist (functionalist) paradigms have in common is precisely that they include, among their many commitments, the adoption of a particular stance. Though the paradigms are short-lived, the associated stances survive their dissolution.

For the *explanatory* task, I think this combined approach is a good start, but incomplete: resources are needed beyond those Boucher and Winther provide (for the *normative* task, I have reservations about stances; Section 5.1). Structuralist (functionalist) paradigms share similarities beyond being motivated by shared stances, and these similarities are essential for explaining why the history of biology looks as it does. What is missing from is something that can both survive radical theory change (as paradigms cannot) as well as elucidate the empirical activities of structuralist (functionalist) biologists (as stances cannot).

Explanatory strategies provide this missing element. By "explanatory strategy" I mean a schema for constructing explanations that partially specifies:

- what constitutes an appropriate target *explanandum*
- what constitutes an appropriate *explanans*

By "partially" in "partially specifies," I mean that the schema must leave out certain key details, such that diverse theories can fill in these details in their own way. By "specifies," I mean that the schema must nonetheless clearly limit what counts as a legitimate way of filling in these details. By filling in details in accordance with this partial specification, the schema is converted into an explanation proper.

Explanatory strategies are independent of stances: one can offer a structuralist (functionalist) explanation without taking a stand on the significance of the *explanandum*. However, insofar as the phenomena foregrounded by structuralists (functionalists) are especially amenable to structuralist (functionalist) explanation, explanatory strategies can help the stance account explain biologists' research practices. Explanatory strategies also allow us to recognize sequences of structuralist (functionalist) paradigms that all favor explanations that realize the same strategy. Explanatory strategies are a reusable resource for constructing paradigms and can outlive them. Finally, explanatory strategies can be integrated by constructing complex explanatory chains incorporating both of them.

It is thus *prima facie* plausible that an analysis of structuralism and functionalism in terms of explanatory strategies will satisfy all five desiderata, while also complementing both stance and paradigm analyses. But it is well known where the devil lurks.

2 Explanatory Strategies

2.1 Structuralism and Functionalism II

In this section, I make the case for analyzing structuralism and functionalism in terms of explanatory strategies. I begin by discussing a range of historical material that any such account must capture, using it to introduce important conceptual clarifications (Sections 2.1–2.4), then make my core case for the importance of explanatory strategies (Sections 2.5–2.6).

The first clarification concerns a common, but misleading, way of presenting the difference between structuralism and functionalism. It is sometimes stated that structuralists and functionalists disagree over whether form is explanatorily prior to function (structuralism) or vice versa (functionalism). For instance, E. S. Russell (1982 [1912], xi) asks, "Is function the mechanical result of form, or is form merely the manifestation of function or activity?" and Stephen Asma (1996, 12) writes, "the question was whether specific organic structure was the result of specific function or vice versa." However, while functionalists do, in an important sense, treat function as prior to form, structuralists do *not* treat form as prior to function *in the corresponding sense.*

The relevant sense of priority here is explanatory priority. In any given explanation, the *explanans* is explanatorily prior to the *explanandum*. To say that functionalists treat function as prior to form is thus to say that functionalists explain form (*explanandum*) by invoking the function served by that form (*explanans*) – and vice versa for structuralists. Note that this relativizes priority to particular explanations: what is *explanans* in one context may be *explanandum* in another.

This adequately captures the functionalist side of the dispute, but it mischaracterizes the structuralist position. This is best appreciated in the light of examples. For functionalists, consider Lamarck and Cuvier. Actually, it is tendentious to call

Lamarck (2011 [1809]) a "functionalist," as his theory is really a hybrid theory involving the interaction of two processes: (a) an orthogenetic tendency for lineages to increase in complexity over time (structuralist) and (b) a mechanism by which novel structures arise in response to new organismal needs (functionalist). For now, consider the latter only. As organisms change their behavior to meet new needs, their physiology changes, leading to structural modifications that are inherited by their descendants. In this way, novel structures arise to fulfill particular functions. For the functionalist portion of Lamarck's theory, structure is *explanandum*, function is *explanans*.

So also for Cuvier (Rudwick 1997), who explained the features of organisms in terms of their conditions of existence (Coleman 1964; Outram 1986; Appel 1987). Cuvier saw organisms as tightly integrated arrangements of anatomical parts, each particular arrangement being determined by the needs associated with an organism's particular form of life (Novick 2019). Once again, structure is *explanandum*, function is *explanans*.

So far, so good. But now consider how Richard Owen, arch-structuralist, attempted to refute these functionalist views. Not only did Owen not argue that structure explains function, his arguments altogether precluded that possibility. Consider Owen's (2007 [1849]) analysis of the bat's wing, the dugong's front fin, and the mole's forelimb (Figure 1). Each is adapted to a different function: the bat's wing for flight, the dugong's fin for swimming, and the mole's forelimb for digging. Nonetheless, each is structurally very similar, consisting of the same bones in the same arrangement. The same basic structure thus serves many functions.

Because of this one-to-many relationship between structure and function, Owen argued that function could not explain structural correspondences. Considering the bat's wing in isolation, one might try to explain how its underlying structure is suited to flying, but that structure's recurrence in the dugong's fin (swimming) and the mole's forelimb (digging) undermines that functional explanation. Moreover, that structure is not necessary for any of those functions, which are achieved by other means in other groups (e.g., insects, fish, and caecilians, respectively). Geoffroy made similar arguments (Appel 1987, 85; Asma 1996, 16).

Owen's argument, however, equally foreclosed the possibility of explaining function in terms of structure. Just as the limbs' functions could not, in virtue of their differences, explain the *sameness* of structure, so too the limbs' shared structure could not explain their *differences* of function. Granted, the differences in structure between the various forelimbs (e.g., the long, thin fingers of the bat compared to the stubby fingers of the mole) might explain their differences in function, but Owen allowed that these modifications of the underlying archetypal pattern were to be explained functionally.

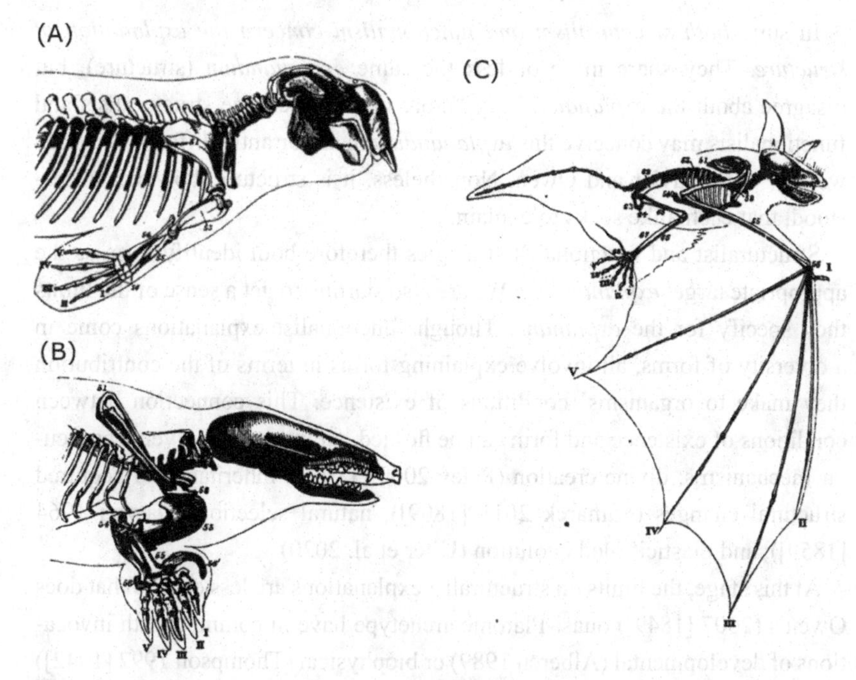

Figure 1 Richard Owen's diagrams of the (A) dugong, (B) mole, and (C) bat forelimbs, showing their structural correspondences

Owen thus did not explain function in terms of deep structure. Indeed, *Owen did not attempt to explain function at all.* Owen was concerned to explain structure, just as Lamarck and Cuvier were. However, unlike Cuvier, he understood structure in terms of two distinct *explananda*, each requiring its own *explanans*: (1) shared archetypal patterns that served distinct functions in distinct groups and (2) modifications to those archetypal patterns for the sake of serving particular functions.

Owen's argument showed that function is not explanatorily prior to structure, but not that (archetypal) structure is explanatorily prior to function. It established structure as requiring explanation *independent* of function; function was relegated to explaining secondary modifications of this stable underlying structure (Boucher 2015, 384). The same holds for the structuralist portion of Lamarck's theory, in which lineages inherently become more complex over time. Here, too, the *explanandum* is form, but the *explanans* is independent of function, grounded instead in the material nature of organism (Newman and Bhat 2011). For both authors, there is a sense in which structure is 'prior' to function: both treat functional modifications as secondary deviations superimposed atop a deep structure that is independent of function. However, this is not *explanatory* priority, and using it encourages a misleading equivocation (see Appel 1987, 2).

In sum, *both structuralism and functionalism concern the explanation of structure.* They share more or less the same *explanandum* (structure), but disagree about the *explanans*. I say "more or less" because structuralists and functionalists may conceive this *explanandum* in importantly different ways, as we saw with Cuvier and Owen. Nonetheless, it is structure, however understood, that each alike seeks to explain.

Structuralist and functionalist strategies therefore both identify form as the appropriate target *explanandum*. We are also starting to get a sense of the limits they specify for the *explanans*. Though functionalist explanations come in a diversity of forms, all involve explaining forms in terms of the contribution they make to organisms' conditions of existence. This connection between conditions of existence and form can be fleshed out in terms of diverse particular mechanisms: divine creation (Paley 2009 [1803]), inheritance of acquired structural changes (Lamarck 2011 [1809]), natural selection (Darwin 1964 [1859]), and plasticity-led evolution (Uller et al. 2020).

At this stage, the limits on structuralist explanations are less clear. What does Owen's (2007 [1849]) quasi-Platonic archetype have in common with invocations of developmental (Alberch 1989) or biophysical (Thompson 1992 [1942]) processes? For now, all we can say is that they all explain form as independent of function. This is a weak limit; it says what structuralist explanations are not, but not what they are.

We therefore still require a positive characterization of structuralist explanations. This will become clearer by considering structure/function disputes arising in evolutionary contexts. By the end, this will reveal the unity underlying both evolutionary and nonevolutionary structuralist explanations, as well as the possibility of explanatory strategies that are neither structuralist nor functionalist (Section 2.5).

2.2 Historicizing Functionalism

Darwin, in his ecumenical fashion, respected the insights of both the partisans of Unity of Type (structuralists) and the partisans of Conditions of Existence (functionalists). Then, in his quietly polemical fashion, he subordinated the former to the latter. In doing so, he developed a sophisticated response to Owen's challenge and reconfigured the nature of structure/function disputes.

Owen had shown that function could not explain deep structural correspondences. Recognizing the force of his argument, Darwin (1964 [1859], 206) accommodated it:

> It is generally acknowledged that all organic beings have been formed on two great laws – Unity of Type, and the Conditions of Existence. By unity of type

> is meant that fundamental agreement in structure, which we see in organic beings of the same class, and which is quite independent of their habits of life. On my theory, unity of type is explained by unity of descent. The expression of conditions of existence, so often insisted on by the illustrious Cuvier, is fully embraced by the principle of natural selection. For natural selection acts by either now adapting the varying parts of each being to its organic and inorganic conditions of life; or by having adapted them during long-past periods of time. [. . .] Hence, in fact, the law of the Conditions of Existence is the higher law; as it includes, through the inheritance of former adaptations, that of Unity of Type.

Darwin made two crucial moves in this passage. First, he explained Unity of Type in terms of unity of descent. Dugongs, bats, and moles share structurally similar forelimbs because they inherited those limbs from an ancestor whose forelimbs bore that structure. Natural selection caused this structure to diverge in different lineages to serve different functions (satisfying Conditions of Existence). Thus far, Darwin's argument seems to grant Owen's point entirely. Selection superimposes minor functional modifications atop inherited structures, fine-tuning them and no more.

Enter the second move: Darwin's argument that Conditions of Existence is the "higher law," because these inherited structures are themselves "former adaptations." Darwin thereby subordinated Unity of Type to Conditions of Existence. Yes, selection superimposes modifications on top of inherited structure, but inherited structure is itself the product of past selection. Unity of Type is the byproduct of Conditions of Existence operating over time, rather than at a single moment of creation.

By introducing history, Darwin freed functionalism from assumptions of optimality – an insight that dawned only gradually (Ospovat 1981), even after he'd recognized roles for both "direct adaptation" and "hereditary taint" (Barrett et al. 2009, B46). Darwinian explanations of current structure require a two-part *explanans*, covering both current and past functions. Past functions explain ancestral structure; current function explains modifications to that inherited structure. Explanations are thus functionalist all the way down. Though current function alone cannot explain deep structural correspondences, Darwin had at his disposal an entire historical sequence of functions. Thus, the one-to-many relationship between structure and current function presents no conceptual difficulty to the Darwinian functionalist.

On one level, this innovation marked a significant break, introducing a novel explanatory resource (past function) into the functionalist's arsenal, thereby defanging one of the most potent structuralist arguments. On another level, however, Darwin left matters largely unchanged. The central issue in structure/function disputes was still the role of function in explaining structure. Darwin developed a powerful new type of functionalist explanation, forcing subsequent structuralist critics to seek out its limits. But, though the details changed, the fundamental explanatory task remained the same: show either that function explains structure or that something else does.

2.3 The Generation and Spread of Variation I

Though Darwin did not change the basic nature of structure/function disputes, Darwinian theorizing did eventually bring into focus a distinction that will help us develop a positive characterization of structuralist explanations. I have in mind the distinction between the *generation* and the *spread* of variation (Love forthcoming, sec. 1.2). In evolutionary contexts, structure/function disputes primarily concern the generation of variation, and only secondarily the question of how variants spread. While this point cannot be *directly* extended to structure/function disputes occurring in non-transmutationist contexts, an indirect extension is possible (Section 2.5).

The distinction between the generation of variants and the spread of variants within a population is central to contemporary evolutionary theory (Endler 1986). This distinction emerged in the context of early population genetic modeling (e.g., Dobzhansky 1982 [1937], 13). These models incorporated two kinds of causes: causes that introduce genetic variation into a population (mutation, migration) and causes that determine which variants spread through that population (selection, drift). While the foregoing presentation emphasizes genetic causes, the point is general: evolution via nongenetic inheritance requires that those variants both be generated and spread through the population (see Lu and Bourrat 2018).

This distinction is illuminating wherever transmutation is a live theoretical option, even for biologists who do not accept it explicitly. Indeed, if transmutationist theorists do not make explicit use of distinction, this is often precisely because they present mechanisms for the generation of variation that also account for its spread – that is, they solve both problems in a single step. For example, both halves of Lamarck's (2011 [1809], chaps. I.VI–VII) theory invoke sources of variation that act uniformly across populations, obviating issues of spread; a similar point can be made for Thompson's (1992 [1942]) explanation of the role of physical forces. Goldschmidt (1982 [1940])'s solution has a different structure: his "hopeful monsters" are, if viable, so different from other individuals of their species that they found a new population in a single step. In that new population, they are *ipso facto* immediately fixed.

Even considering a species fixist such as Cuvier, the distinction is helpful. Cuvier's anti-transmutationist arguments, interpreted in this light, aim to show that, even if variants arise that modify a type, those variants will not be viable and will not spread. Thus, in explaining form, only the problem of the generation arises. While this may not seem illuminating now, it points to the fact (to be argued for in Section 2.5) that how variation is generated is more fundamental to structure/function disputes than how it is generated. Since the problem of spread is restricted to transmutationist contexts, while the problem of generation is not, this will help us see why structure/function disputes survived the rise of evolutionary theory.

Thus, though the distinction between the generation and spread of variation is most closely associated with functionalists, historically, it can illuminate the views of structuralist and functionalists alike. Both functionalists and structuralists can solve both problems together; both structuralists and functionalists can keep the problems separate. In relying on this distinction, then, I am not prejudging matters in favor of either camp.

2.4 The Generation and Spread of Variation II

We are now ready to revisit the question of what distinguishes structuralists and functionalists. In this section, I defend two points. First, structure/function disputes, in evolutionary contexts, concern the sources of direction in evolution. Second, these disputes are driven by disagreements concerning how variation is generated. While these may generate downstream disagreements about how variation spreads in a population, *structure/function disputes fundamentally rest on competing views about how variation is generated.*

Let us start, not with conflict, but rather with a biologist who integrated the two approaches to at least some degree. Charles Darwin (1964 [1859], chap. 5) devoted an entire chapter of the *Origin* to what he called "laws of growth." He was especially interested in correlated variation, in which two seemingly independent parts (call them S, for *selected*, and C, for *correlated*) vary together. When variation in S is correlated with variation in C, selection for changes in S will result in changes to C. To explain changes to C, we must invoke both function (selection acting on variation in S) and facts about the developmental system that explain why variation in S is correlated with variation in C.

In this explanation, the direction of evolutionary change is explained in part by selection, in part by potentialities of the developmental system. Though this is a happy case of integration, attempting to understand the sources of direction in evolutionary change is, in other cases, precisely what gives rise to structure/function disputes. Among the central claims of the (functionalist) modern synthesis was that selection is the *only* direction-giving factor in evolution (Mayr 1969; Mayr and Provine 1981). This claim was intended to exclude both (a) neo-"Lamarckian" functionalist theories that let functional needs influence the generation of variation and (b) structuralist theories that dispensed altogether with the need for selection to explain directional evolutionary change (e.g., Schindewolf 1993 [1950]).

The claim that selection is the only direction-giving factor in evolution requires careful exegesis. If taken to imply that the generation of variation plays *no* role in determining the direction of evolutionary change, it is remarkably strong. It is strong because it is committed to the view that variation is *isotropic* (or nearly so): that variation is generated equally in all directions (cf. Minchin in the epigraph). On this view, the processes that generate variation furnish selection with variants of all sorts, and all the work of determining which variants survive into future generations is done by selection (or by drift, which, being random, is *ipso facto* not direction-giving). By contrast, *any* failure of isotropy leaves room for the generation of variation to affect the direction of evolutionary change, even within standard population genetic models (Arthur 2001; Fusco 2015).

The assumption that variation is isotropic has two problems: it is in most cases incoherent, and where coherent it is frequently false (Salazar-Ciudad 2021). It is incoherent when the variants in question are considered qualitatively. For example, at some point in the evolution of birds, scales evolved into feathers (Wagner 2014, chap. 9). But scales might have evolved into sundry other things: skin, an outer shell, or any number of unnamed outer coverings that exist only in possibility space. Could there be variation in all of these indefinitely many directions? Not in any actual, finite population.

By contrast, where a property is quantitative, allowing variation to be characterized in terms of changes in a limited number of directions (e.g., height can only increase or decrease), the assumption that variation is isotropic is coherent. Coherent, yes, but still routinely violated (Arthur 2001). For a simple illustration, consider a domesticated species that has been pushed by artificial selection to an extreme value for some trait – "extreme" in the sense that breeders struggle to push the trait any further in the same direction. The difficulty is precisely that variation is nonisotropic: nearly all new variants for the trait are in the direction of its original value (see Darwin 1964 [1859], chap. 1 on pigeon fancying).

But this strong form of the assumption cannot be what the architects of the synthesis actually meant (contra Pigliucci 2019, 17), for the simple reason that they were aware that variation is rarely truly isotropic (Svensson 2021). For instance, in the case of correlated variation just considered, both structural and functional factors contributed to evolutionary direction. Without selection favoring changes in S, there would be no change in either S or C. But selection alone cannot explain the directional change in C. The features of the developmental system that produce correlated variation must be invoked to explain this. If S and C were correlated in the inverse direction, the same selection pressure on S would have driven C to evolve in the opposite direction. So features of development are direction-giving factors.

Read more charitably, the claim that selection is the only direction-giving factor in evolution is really three-fold: it claims (1) that variation is present in *many* directions, (2) that biases affecting the generation of variation are generally weak, local, and overpowered by selection, and (3) that selection is *necessary* for directional evolutionary change. The first assumption ensures that selection will have ample material to choose among. The second assumption allows the synthesist to acknowledge the ineluctable role of biased variation while minimizing its force. In saying that such biases are weak and local, the claim is that they change as lineages evolve and so cannot explain long-term directional trends. The third assumption denies that any factor other than selection (structuralist or otherwise) is sufficient to explain such long-term directional change without selection's aid. Together, these assumptions rule out the views rejected by the synthesis: orthogenetic invocations of "mutation pressure" (which explain long-term directional change without selection), Goldschmidt's saltationist account of "hopeful monsters" (which allows selection only a bit role in eliminating hopeless monsters), and neo-Lamarckian views on which variation is generated nonrandomly with respect to function (rendering selection redundant). What remains is neo-Darwinian gradualism: directional evolution results from iterated rounds of selection accumulating small changes over time.

We are now ready to consider my second claim: that structure/function disputes, in evolutionary contexts, fundamentally involve disagreements about the generation of variation. As we have seen, the claim that natural selection is the only direction-giving factor in evolution rests on claims about how variation is generated: namely, that the processes that generate variation do not do so in ways that determine the direction of evolutionary change without the need for selection. The synthetic claim is true only if neo-Lamarckian, orthogenetic, and Goldschmidtian accounts of the generation of variation are false.

The dispute between Minchin and Thompson over the spicules of sponges likewise turns on a claim about how variants are generated. Minchin explains the evolution of spicules in terms of selection acting on "variation in every direction." Selection is of especial importance because variation by itself is nondirectional. Thompson, by contrast, allows the generation of variation to determine form, because this generation is determined by physical forces acting (uniformly) on the spicules of any given species. No role remains for selection.

Thompson himself rightly identified this disagreement about how spicule variants are generated as "the fundamental difference" between Minchin's functionalist approach and his own structuralist approach (see epigraph).

One final example suffices to make the point. Stuart Newman and Ramray Bhat (Newman and Bhat 2008, 2009) argue that the diversification of early animal forms was largely the result of "dynamical patterning modules." These modules involve deeply conserved toolkit genes whose protein products affect the physical processes in which cells engage (e.g., cell–cell adhesion, diffusion). On their view, these physical processes tightly constrain possible form: for instance, slight changes in adhesive properties generate large and predictable changes in the form of mass of adhering cells. They contrast their approach with the "neo-neo-Darwinian" approach of evo-devo biologists such as Sean Carroll (2008), according to which "unusually intense selection on [regulatory] DNA ... led to extremely rapid, but still incremental, diversification of form during a narrow period of time at the Precambrian-Cambrian boundary" (Newman and Bhat 2008, 2). At issue, once again, is the generation of variation: the "neo-neo-Darwinian" hypothesis assumes the existence of incremental variants; Newman and Bhat's hypothesis denies it.

Thus, structure/function disputes, in evolutionary contexts, rest on disagreements about how variants are generated. This may result in downstream disagreements about how variants spread, but such disputes are secondary.

2.5 Structuralism and Functionalism III

Once we recognize that the problem of the generation of variation is at the core of (evolutionary) structure/function disputes, three major points follow. The first concerns the possibility of generalizing this approach to nonevolutionary views. The second concerns the varieties of structuralist and functionalist explanations. The third concerns the possibility of explanations of form that are neither functionalist nor structuralist. I take these up in turn. In the process, I shall offer a precise characterization of the nature of both structuralist and functionalist explanatory strategies.

The distinction between the generation and spread of variation is sensible only in contexts where transmutation is a live theoretical possibility. However, the centrality of the problem of the generation of variation can, with suitable modification, be generalized even to nonevolutionary structure/function disputes. The core idea is that study of the generation of variation is just a specifically *evolutionary* way of studying broader questions about what forms are *possible* (what forms can be produced at all) in terms of what forms are *accessible* (what forms can be reached from a given starting point).

In Section 2.1, structuralism was given a spare, negative characterization: structuralists argue that form is *independent* of function. We can now start to give a positive characterization. Structuralists are committed to the idea that the space of possible forms has internal *structure*, and that this structure is (a) explicable independent of function and (b) important to explaining the actual forms seen in the world around us.

In evolutionary contexts, the structure of the space of possible forms involves both constraints (certain forms are difficult or impossible to generate) and drive (certain forms are especially likely to be generated). Constraints and drive are necessary concomitants (Arthur 2001; Salazar-Ciudad 2021), and are collectively referred to as "biases." Biases can in turn be explained in terms of underlying generative processes. For instance, Newman and Bhat (2008) explain the evolution of early metazoan form in terms of the range of body plans that can be generated given the physics of cell–cell adhesion (*inter alia*).

Bias, however, is just a specifically evolutionary way of making sense of the idea that the structure of the space of possible forms is *explicable*. The structure of the space of possible forms is thus not a mere reification of the range of forms actually seen. Rather, that structure can itself be traced to some underlying cause. Claims that certain forms are (im)possible or (un)likely are not absolute: they are *conditional* on that underlying cause. Newman and Bhat's invocation of the physics of adhesion is a conditional claim of this sort: *given* that cells are adhering to one another in specific ways, only a specific, limited range of forms can be generated.

Possibility claims made by nonevolutionary structuralists are conditional in just the same way. Consider Owen's vertebrate archetype. *Given* that a dugong is a vertebrate, its limb is constrained to possess a certain set of bones in a certain arrangement. Within this constraint, the limb can be reconfigured to serve its swimming function, but limb forms that do not fit the archetypal pattern are off the table, *even if they might be superior at serving that function*. Of course, one might ask what the archetype is and how it is able to constrain form in this way (a thorny matter; Camardi 2001), but we can recognize the structuralist nature of Owen's explanation without having an answer.

In light of this, I offer the following characterization of the basic structuralist and functionalist explanatory strategies (Table 1):

- **The functionalist strategy** specifies form as *explanandum* and function as *explanans*. The presence of particular aspects of form is to be explained by their contribution to meeting organisms' needs (note: the relevant organisms may be ancestral to the organisms whose form is being explained).

Table 1 Structuralist and functionalist explanatory strategies. Further explanation in text

	Explanandum	**Abstract *Explanans***	**Concrete *Explanans***
Functionalist Explanations	Form	Functional needs determine form. A concrete *explanans* is generated by specifying the *causal link* between function and form	[1] Divine creator considers species needs (Paley) [2] Environmental changes generate functional variants (Lamarck, Jablonka) [3] Natural selection favors functional variants (Darwin, Dobzhansky)
Structuralist Explanations	Form	The structure of the space possible form determines (actual) form. A concrete *explanans* is generated by specifying the *source* of this structure.	[1] Archetype constrains form (Goethe, Owen) [2] Biophysical processes constrain form (Lamarck, Thompson, Newman) [3] Developmental processes constrain form (Goldschmidt, Alberch)

- **The structuralist strategy** specifies form as *explanandum* and the structure of the space of possible form as *explanans*. The structure of the space of possible forms must be *explicable* in terms of some underlying cause, and this underlying cause must be *independent* of function.

Both furnish recipes for generating explanations of form, in accordance with the description of explanatory strategies given at the outset (Section 1.3). Each specifies what an appropriate *explanans* must involve, but these specifications are partial: they leave room to fill in details in accordance with particular theoretical commitments. Partial, but not *empty*: they set meaningful limits to what counts as a legitimate realization of the strategy. These limits deserve further elaboration.

The functionalist strategy requires that form be explained in terms of function. That is, forms must be accounted for in terms of how they contribute to fulfilling organisms' needs. However, this connection can be made in many ways. Perhaps a divine intelligence foresees the needs of organisms at the moment of creation (Paley 2009 [1803]), or perhaps natural selection accumulates favorable variants over countless generations (Darwin 1964 [1859]). Perhaps it is only the needs of a single species that matter (Cuvier; see Rudwick 1997), or perhaps we must consider the needs of an entire lineage spanning a long range of time (Darwin 1964 [1859] again). In transmutationist contexts, the needs of organisms may influence what variants spread through a population (Dobzhansky 1982 [1937]; Carroll 2008), but it also might directly influence which variants are generated (Lamarck 2011 [1809]; Laland et al. 2015).

The structuralist strategy requires that form be explained in terms of the structure of the space of possible form, which renders certain forms likely and others unlikely (or, in the extreme case, impossible). This structure must itself be explicable in terms of some underlying cause, and this cause must be independent of function. As with the functionalist strategy, this specification allows for a wide range of ways of filling in the details (though note that the emphasis on possible form means that structuralist explanations never directly target the spread of variation). A structuralist explanation can focus entirely on the raw physical properties of organic materials (Empedocles; see Irwin and Fine 1996, 87–88), or it may focus on the physical interactions of these materials with their environment (Thompson 1992 [1942]). It may invoke a fixed, possibly idealist archetype (Owen 2007 [1849]), or it may invoke evolved developmental types (Alberch 1989). Finally, it may be gradualist (Wagner 2014), or it may be saltational (Goldschmidt 1982 [1940]; Newman and Bhat 2008).

This brings us to the third and final point of this section: the possibility of explanations that are neither functionalist nor structuralist. This is an immediate consequence of the fact that the structuralist strategy involves more stringent requirements on structuralist explanations than just "not functionalist." Neutralist explanations (Kimura 1983), which assume that variation is produced in many directions, but that the spread of variants is due primarily to drift, furnish a clear example. They are not functionalist, since selection plays no role in explaining form, but neither are they structuralist, since they do not invoke biases affecting the generation of variation.

If this analysis is convincing, it evades a dilemma raised by Boucher (2015, 391). Boucher argues that explanatory strategies either (**horn 1**) tie themselves to particular theoretical commitments that are not accepted by all relevant parties or (**horn 2**) collapse into the value-commitments associated with stances. To escape these "therrble prongs" (Joyce 2012 [1939], 628.5), suitable notions of "function" and "structured possibility space" are required. I think I have provided these, relying on notions of organism's needs and of the conditions under which particular forms can and cannot be generated, respectively, that do not require any specific theoretical commitments. To be sure, these notions are not especially informative by themselves: to generate an *explanation* one needs to fill in a great deal of detail. But neither are they contentless: they pick out clusters of nontrivially similar explanations. The very ability to recognize that neutralist explanations fall outside of both schemata attests to this. Explanatory strategies are thus sufficiently abstract to span the history of biology, yet still sufficiently contentful to identify significant similarities across this history, not just in *what* phenomena biologists seek to explain, but also in *how* they explain them.

2.6 Local Structure/Function Disputes

The foregoing analysis of structuralism and functionalism helps to clarify what is at stake in particular structure/function disputes. As this concerns both the explanatory and the normative aims of my analysis (Section 1.3), I limit my focus to evolutionary contexts, where the normative issues are still live. When evolutionary biologists find themselves caught in such a dispute, what is the source of their disagreement? How can it be resolved? I contend that these conflicts concern competing answers to specific contrastive why-questions. This will help us to understand why the recurrence of such disputes is an ineradicable feature of evolutionary biology (fourth desideratum), yet also how particular disputes can be empirically resolved (third desideratum).

Evolutionary biologists both (i) develop general accounts of evolutionary processes and (ii) explain particular evolutionary outcomes. The former has occupied most of the philosophical attention paid to structuralism and functionalism (e.g., Amundson 2005; Boucher 2015). It also captures how many biologists have seen themselves. Consider once again the epigraph: D'Arcy Wentworth Thompson, in the midst of a specific dispute about the evolution of the spicules of sponges, steps outside those particulars to comment on the quite general difference between physicalist and Darwinian "conception[s] of the causation and determination of Form."

If we take explanatory strategies as the core of our analysis of structuralism and functionalism, however, then we should instead focus first on local disputes, and only later (Sections 4–5) return to more general questions about evolutionary theory as such. In these local disputes, particular structuralist explanations compete with particular functionalist explanations. Below, I will show that how explanations of the evolution of any particular structure can be broken down into answers to distinct why-questions (Van Fraassen 1980, chap. 5). Once the various why-questions involved in a particular dispute are appropriately disentangled, we can distinguish between structuralist and functionalist answers to them, revealing both direct points of conflict and how to empirically resolve them.

Consider the case of digit reduction in amphibians (Alberch and Gale 1985; see Stopper and Wagner 2007; Wiens and Hoverman 2008). Both anurans (frogs) and urodeles (salamanders) contain lineages that have lost all or part of certain digits. Two facts stand out. First, there are trends within both the anuran and urodele lineages, even though digit loss has independently evolved multiple times within both of them. Anurans that have lost only one phalange always lose a phalange from the first digit, while urodeles that have lost only one phalange always lose a phalange from the fourth digit. Second, while both groups show trends in digit loss, they do not show the *same* trend. Anurans and urodeles lose distinct phalanges first.

What explains these facts? Alberch and Gale contrast two strategies. The first is that urodele and anuran developmental systems produce biased digit loss phenotypes, such that, if a mutation causes loss of a single phalange, it always causes loss of a phalange from the first (anurans) or fourth (urodeles) digit. Selection may sometimes favor digit reduction, but (in anurans) it does not favor loss of a phalange from the first digit over loss of a phalange from the third digit – it never sees the latter. Digit loss trends are thus explained in terms of developmental systems biasing the generation of variation.

The second strategy is functionalist. It regards developmental biases as both weak (many different digit loss phenotypes can be produced) and local (restricted to lower ranks of the taxonomic hierarchy). Given these assumptions, shared biases cannot explain these evolutionary trends; selection is (probably) required. On this view, trends seen in multiple lineages are the result of similar selection pressures operating in each lineage.

This neatly fits the account I have been developing: the functionalist assumes that variation is sufficiently close isotropic to render selection the major determinant of evolutionary direction, whereas the structuralist assumes that variation is sufficiently anisotropic to render developmental bias the major determinant of evolutionary direction. In moving to a particular case, however, we can ask more precise questions. The task is not to explain form in general; it is to explain the evolution of particular forms.

Explanations answer contrastive why-questions: "why X rather than Y?" Why, for instance, did anuran lineages lose digit one first (if they lost digits at all), rather than any other digit? By analyzing the question "what explains amphibian digit loss trends?" in terms of a collection of more specific questions in this contrastive form, we can precisely localize the structure/function dispute at issue. Each such specific question can receive either a structuralist or a functionalist answer; these can be characterized as follows:

- **Functionalist answers** explain why X evolved, rather than Y, by invoking either (a) selection favoring the spread of X over Y or (b) functions that bias the generation of variation in favor of X over Y.
- **Structuralist answers** explain why X evolved, rather than Y, by invoking generative processes that bias the generation of variation in the direction of X (as compared to Y).

There may also be neutralist answers (X, rather than Y, happened to be fixed by drift), which are neither structuralist nor functionalist, but these are not relevant here. I also set aside (b)-type functionalist answers. While proponents of the extended evolutionary synthesis (Laland et al. 2015) offer functionalist explanations of this type, Alberch and Gale focus on (a)-type functionalist answers, and I shall follow them.

There are many why-questions one might ask about amphibian digit loss. Here is a nonexhaustive sample:

(W1) Why did these anuran lineages lose a phalange from their first digit, while those other anuran lineages did not?

(W2) Why did these anuran lineages lose a phalange from their first digit, rather than gain (a) phalange(s) or not change at all?

(W3) Given that they all lost a phalange, why did these anuran lineages all lose a phalange from their first digit, rather than some other phalange?

(W4) Given that they all lost a phalange, why do anuran and urodele lineages lose a different phalange first, rather than the same phalange?

(W5) Why did this particular anuran lineage lose a phalange from its first digit, rather than from some other digit?

Despite their structuralism, Alberch and Gale offer functionalist answers to both W1 and W2. They accept that there was likely selection pressure for digit reduction (W2) that operated only in certain lineages (W1). Note that both questions could be given structuralist answers. For W2, for instance, Alberch and Gale might have argued that amphibian developmental systems can only generate variation in the direction of reduction, never increase (had they done so, they would have been wrong; Borkin and Pikulik 1986).

Conversely, in allowing that a functionalist can accept local developmental biases, Alberch and Gale acknowledge that said functionalist could say, of a particular lineage, that developmental bias at least partially explains why that lineage lost a particular phalange first. The functionalist, that is, could offer a structuralist answer to W5, at least for some lineages.

The conflict between Alberch and Gale and their hypothetical functionalist opponent thus concerns only questions W3 and W4. W3 asks, over a range of cases, why evolution went in X direction rather than Y direction, while W4 asks why it went in X direction (rather than Y) in one set of cases and in Y direction (rather than X) in another. In both cases, the structuralist answers in terms of biases affecting the generation of variation, whereas the functionalist assumes that variation was generated in both directions, with selection choosing between them.

Centering local structure/function disputes in this way has three effects. First, it allows us to precisely localize the conflict, identifying disagreement over how to answer particular questions. It thereby helps to identify the information required to resolve the dispute empirically. The central disagreement turns on how and to what extent the anuran and urodele developmental systems bias the generation of digit loss phenotypes; this, then, is the information needed to resolve it. While it may be practically difficult to obtain it, there is no conceptual difficulty. Indeed, Alberch and Gale's paper presents experiments designed to provide just this information. They disrupted limb development, looking for patterns in the digit-loss phenotypes generated thereby. They found a close match between evolved and experimentally generated phenotypes, and took this to support their structuralist answers to W3 and W4.

Second, while focusing on explanatory strategies clarifies what is at stake in particular structure/function disputes, it renders the appellations "structuralist" and "functionalist" rather fuzzy. In this dispute, both camps agreed that functionalist answers are required for some why-questions (W1, W2) and that structuralist answers are required for others (W5). If we keep our focus on explanatory strategies, then Alberch and Gale are "structuralists" only because they give structuralist answers to more why-questions than their "functionalist" opponents. This is a weak, context-relative sense of being a "structuralist": if they were instead debating someone who offered structuralist answers to W1 and W2, they would come out as "functionalists" instead. If "structuralist" and "functionalist" are to designate intrinsic, stable identities of particular biologists, we must look elsewhere. I defer further consideration of this issue until later (Section 4–5.1).

Third, my analysis explains why structure/function disputes are an ineradicable part of evolutionary biology. Anyone who allows that variation is anisotropic (that is, everyone) must allow that some why-questions will receive structuralist answers. Anyone who accepts that natural selection plays a role in determining which variants spread through a population (that is, everyone) must allow that some why-questions will receive functionalist answers. To accept evolutionary theory is *eo ipso* to accept the legitimacy of both explanatory strategies.

At the same time, this leaves plenty of room for disagreement over which particular why-questions require which particular kind of answer. And, since (a) actually answering the relevant contrastive why-questions is frequently rather difficult and (b) evolutionary processes themselves evolve (precluding a one-size-fits-all approach to evolution), there will always be new opportunities for disagreements. Every new why-question is a potential locus for a structure/function dispute; every new theoretical advance is a chance to re-open old disputes. Any given structure/function dispute can, with sufficient luck and ingenuity, be resolved, but the structure/function dispute as such is an ineliminable feature of evolutionary biology.

2.7 Internalism and Externalism

I have analyzed Alberch and Gale's work on amphibian digit loss as an instance of a structure/function dispute. This is not, however, how Alberch himself treated it. Rather, Alberch (1989) referred to his views as "internalist," contrasting them with those of his "externalist" opponents. This terminology is common (e.g., McShea 1991; Sterelny 2000; Arthur 2015; Fábregas-Tejeda and Vergara-Silva 2018; Stoltzfus 2021; Brown 2022). It is most often used specifically to capture the difference between evo-devo and "neo-Darwinian" evolutionary theory, though not always (Sansom 2008 extends it back to Darwin). On the surface, the distinction between internalism and externalism looks very much like that between structuralism and functionalism. Thus, for instance, developmental bias is an "internal" factor in evolution, whereas selection is an "external" factor. The relationship between the two distinctions therefore deserves detailed consideration. I will show that internalism cannot be identified with structuralism, nor externalism with functionalism – the two are orthogonal distinctions.

An internalist explanation selects some aspect of form as its *explanandum* and invokes some cause internal to the organism as the *explanans*. For instance, developmental bias depends on the nature of an organism's developmental system, and so is an internal cause. Orthogenetic theories (e.g., Schindewolf 1993 [1950]), which posit an environment-independent directionality to evolutionary change within particular lineages, are likewise internalist. So are theories emphasizing the role of mutation pressure (Stoltzfus and Cable 2014). When Richard Owen developed an evolutionary version of his archetype theory (Camardi 2001; Rupke 2009), this, too, was internalist.

By contrast, an externalist explanation selects some aspect of organismic form as its *explanandum* and invokes some cause external to the organism as the *explanans*. Explanations in terms of natural selection are paradigmatic externalist explanations because they focus on organism-environment relations: no trait is fit as such, but only in relation to some range of environments (but see below). So far as I can tell, in discussions of internalist/externalist disputes, the externalist is always offering selective explanations.

The internalist/externalist distinction is nonequivalent to the structuralist/functionalist distinction on both sides: there are internalist functionalist explanations, and there are externalist structuralist explanations. For the former, there is a long history of functionalist, selection-based explanations that invoke selection pressures generated within the organism (Whyte 1960; Schank and Wimsatt 1986; Wagner and Schwenk 2000; Wimsatt 2013; Novick 2019). In these cases, the source of selection pressure is not organism-environment fit, but rather the relationship between different parts of the organism. Where parts of the organism are tightly functionally integrated, the specifics of the environment may have little to do with which changes are viable. In such cases, the selection involved is purifying selection (resulting in stabilizing selection), rather than the directional selection that is most prominent in externalist explanations.

On the other side, structuralist explanations can essentially involve factors external to the organism. For example, explanations of structure that focus on physical forces often depend on the particular medium in which an organism lives. The laws of physics may hold universally, but different forces dominate in different environments (Thompson 1992 [1942]). Thus, the physical determination of form is hardly purely internal. Moreover, even developmental processes need not be purely internal: development is often highly environment-sensitive, as students of developmental plasticity (West-Eberhard 2003) and "eco-devo" (Sultan 2015; Gilbert 2016) note.

The structuralist/functionalist and internalist/externalist distinctions are thus thoroughly cross-cutting. There is also a second, subtler difference between them. The distinction between structuralist and functionalist explanations is a distinction between discrete kinds of explanations. Though they can be integrated in complex explanatory chains that contain both structuralist and functionalist parts (Section 3), the parts themselves remain either structuralist or functionalist (or neither); there is no middle ground. By contrast, the distinction between internalist and externalist explanations identifies the poles of a continuum of environment-dependence. Eco-devo explanations, for instance, essentially depend on interactions between external environmental inputs and internal developmental processes. Thus, the two distinctions must not be conflated.

Of the two, I find that the structuralist/functionalist distinction is usually the more illuminating. Alberch and Gale (1985) disagree with their hypothetical opponents both about how digit loss variation is generated *and* about the role played by the environment in digit loss evolution, but the former disagreement is more fundamental. The reduced role for the environment in their preferred explanation is a *consequence* of the restricted range of digit loss phenotypes available to selection. Since a comparably reduced environmental role can be achieved equally via functionalist means (e.g., strong, environment-independent purifying selection), Alberch and Gale's critique of selection-based explanations of digit loss trends is more distinctively structuralist than internalist.

2.8 Interim Conclusion I

This completes the first stage of my analysis of structuralism and functionalism in terms of explanatory strategies. I have provided a precise characterization of both strategies, and I have shown how, by focusing on these strategies (rather than on paradigms or stances), the first four desiderata identified at the outset (Section 1.3) can be satisfied. My account captures the full historical diversity of structuralist and functionalist biology (1) without rendering structuralism and functionalism so vague as to be explanatorily inert (2). It explains why structure/function disputes are a recurrent feature of the history of biology (3) while nonetheless showing how particular such disputes are amenable to empirical resolution (4).

The second stage is to provide an account of how the two explanatory strategies are related. This is the task of the next two sections. One way the two strategies might be related is by being *integrated* into complex explanatory chains, chains that are therefore neither purely structuralist nor purely functionalist (fifth desideratum). I will show that such explanatory chains play a central role in contemporary evo-devo (Section 3). Another way the two strategies might be related is by one strategy being *subordinate* to the other. I will argue that neither strategy is generally subordinate to the other: where subordination occurs, it is always local (Section 4).

3 Understanding Evo-Devo

3.1 Received Views of Evo-Devo

Thus far, I have considered structuralism and functionalism as sources of conflict in the history of biology. However, understanding structuralism and functionalism in terms of explanatory strategies allows us to see beyond conflict, revealing the possibility of integrated explanations that possess greater explanatory power than either strategy on its own. While the appeal of focusing on structuralists' and functionalists' dramatic disagreements is obvious, it is a mistake to treat antagonism as the default relationship between the two. Integration is at least comparably important. It plays an especially central role in evolutionary-developmental biology (evo-devo), which will be my focus here.

As the name suggests, evo-devo focuses on the relationship between "the two great creative processes of biology" (Arthur 2021, 1): development, which shapes organismic form on the timescale of an individual lifespan, and evolution, which shapes organismic form on the timescale of generations. Evo-devo examines this relationship from both directions, asking how evolution shapes development, as well as how development constrains and channels evolution. Prior to the 1980s (when evo-devo first emerged), development was relegated to the sidelines of evolutionary theory. Bringing it back into the fold involved a fair share of controversy (Amundson 2005).

Much of this controversy centers around the idea that evo-devo presents a specifically structuralist challenge to the functionalist mainstream of evolutionary theorizing. This characterization is endorsed by both biologists (Alberch 1989; Reeve and Sherman 1993; Hall 2000; Jenner 2006; Linde-Medina 2010; Wagner 2014; Svensson 2021) and philosophers (Amundson 1994, 2005; Boucher 2015; Austin 2017; Brown 2022) alike, including both evo-devo's defenders and its critics. Even for those seeking to demonstrate the compatibility of evo-devo and mainstream theorizing, the very need to demonstrate this compatibility turns on evo-devo's structuralist status (Lewens 2009; Brigandt 2021).

This presentation of evo-devo as a structuralist research program highlights its reliance on two ideas: developmental biases and types (Amundson 2005). The basic idea is that there are deeply conserved developmental types (e.g., the "anuran limb"; Alberch and Gale 1985) that bias what variants a particular developmental system can generate. These biases can then be invoked to explain directional evolutionary trends (Alberch 1989; Arthur 2001, 2004), deep conservation of animal body plans (Davidson and Erwin 2006; Erwin and Valentine 2013; Erwin 2020), and patterns relating to the origin of novel body parts (Wagner 2014), *inter alia*.

Why should this lead to conflict between evo-devo and mainstream theorizing? Such conflict has followed evo-devo since its origins in the 1980s, fanned both by those sympathetic to evo-devo (Alberch 1989; Amundson 2005; Wagner 2014) and by those concerned to challenge the challenger (Reeve and Sherman 1993; Coyne 2006; Jenner 2006; Hoekstra and Coyne 2007). Amundson (2005) has suggested that evo-devo's structuralism generates various points of incommensurability with the functionalist mainstream, including: (i) disagreement over whether developmental types have any place in the ontology of evolutionary theory, (ii) reliance on different notions of "constraint," with associated differences in methodology and explanatory standards (Amundson 1994), and (iii) emphasis on distinct explanatory targets (form and change, respectively).

Can these conflicts be resolved? Amundson's (2005, chap. 11) analysis ends at an impasse, seeing no way forward without one or the other side making serious concessions. Much of the work in Amundson's wake, however, considers how evo-devo can make peace with mainstream theorizing, including specifically through the integration of functionalist and structuralist approaches (Breuker, Debat, and Klingenberg 2006; Laubichler 2009; Lewens 2009; Brigandt and Love 2010, 2012; Winther 2015; Amundson 2021; Brown 2022).

What follows is my own contribution to these integrative efforts. My central underlying conviction is that, while it is correct to note the importance of structuralist reasoning in evo-devo, it is mistaken to treat evo-devo as exclusively or even predominantly structuralist. Much of the most powerful work within evo-devo is essentially integrative. Its central explanatory accomplishments frequently involve explanatory chains that obtain their power precisely by integrating structuralist and functionalist explanatory strategies. I will focus on three types of explanatory integration in evo-devo: the use of structuralist considerations to deepen functionalist explanations (Section 3.2), the use of functionalist reasoning to explain the conservation of biases (Section 3.3), and the threefold integration of structuralism, functionalism, and neutralism (Section 3.4).

3.2 Integration I: Using Structuralism to Deepen Functionalism

One way to integrate the two explanatory strategies is to use structuralist considerations to help identify the proper targets of functionalist explanation. Integrative explanations of this type substantially predate the rise of evo-devo. As we have already seen, Darwin (1964 [1859], chap. 5) himself offered a lengthy discussion of correlated variation, one aim of which was to explain why apparently separate traits might not be independently optimizable. Though Darwin self-consciously did not know what explained such correlations, he expected that they were ultimately explicable in terms of as-yet-unknown "laws of growth." Recognizing correlated growth was essential to offering selective explanations, as failure to attend to such correlations would lead one to seek selective explanations in the wrong place.

In the twentieth century, one of the most famous critiques of functionalist biology was, in effect, an extended reflection on laws of growth. I mean, of course, Gould and Lewontin's (1979) classic paper on spandrels. Their basic critique of the "Panglossian paradigm" concerns issues of trait individuation – that is, which features of organisms are capable of varying and evolving separately. The Panglossian paradigm treats individual structures (or features of structures) as independently optimizable. Gould and Lewontin challenge this assumption, arguing that the parts of organisms are developmentally entangled in complex ways, and these patterns of entanglement must be understood in order to properly identify which traits are actually being shaped directly by selection.

The spandrels paper thus critiques, not functionalist reasoning as such, but rather functionalist reasoning that incorrectly individuates traits. If traits are mis-individuated, then one will offer functionalist answers to particular why-questions that actually require structuralist answers. However, the presumption is still that, *once traits are correctly individuated*, functional concerns will play a substantial role in explaining their evolution. The role of structuralist concerns about trait individuation here is, to put it somewhat cheekily, *to deepen functionalist explanations*. "The Spandrels of San Marco," received as a classic of structuralist thought, really urges the integration of structuralist and functionalist explanatory strategies. Moreover, the ultimate benefit of this integration is to properly identify the targets of functionalist explanation.

While the Spandrels paper predates the emergence of evo-devo, a similar type of integration shows up in evo-devo proper. Much of Günter Wagner's (1989, 2007, 2014) work on character identity, for instance, attempts to individuate characters based on their patterns of variability. Very roughly, the idea is that characters are more individuated to the extent to which they are able to vary independently. For example, in many insects, the fore- and hindwings are highly similar, and mutations that affect the one tend to affect the other as well. Dipterans, however, have broken this correlation: instead of hindwings, they possess small sensory organs called halteres which little resemble their forewings. The forewings and halteres of Dipterans are thus more highly individuated than the fore- and hindwings of most insects (Wagner 2014, chap. 2). While Wagner puts this work to properly structuralist ends (Section 3.3), it also serves functionalist ends: individuating characters is essential to identifying targets of selection.

A second form of integration shows up in one of evo-devo's central contentions, the "*cis*-regulatory hypothesis." This hypothesis posits that morphological evolution is more likely to occur via changes in regulatory sequences than in protein-coding sequences (Stern 2000; Carroll 2005; Wray 2007). The basic reasoning runs as follows. Protein-coding genes that affect form are widely reused during development. This applies to both regulatory genes (those whose proteins regulate the expression of other genes) and structural genes (those used to actually build morphological structures). Accordingly, changes affecting protein-coding sequences are likely to have highly pleiotropic effects, and thus to be deleterious: even if advantageous in one area, they are likely to cause problems somewhere else. By contrast, regulatory sequences (e.g., transcription factor binding sites) have more modular effects, generally controlling the expression of one gene in one location. Mutations to such sequences thus offer more fine-grained phenotypic variants to selection, and so are more likely to spread.

Unlike the first form of integration considered, trait individuation is not at issue here. Rather, the point is that the genetic architecture underlying development is structured: mutations affecting some parts of that architecture (*cis*-regulatory sequences) are more likely to be selected than mutations affecting other parts (protein-coding sequences). Taking for granted that selection predominantly favors mutations with small effect sizes, evo-devo biologists argue that such mutations occur nonrandomly with respect to the developmental system. This reasoning is structuralist – it concerns what modifications to form are possible – but it serves functionalist ends, helping us understand how selection modifies development.

A third type of integration that fits (uneasily) under the heading of using structuralist considerations to deepen functionalist explanations involves the incorporation of developmental bias into mathematical population genetic models. In a sizable body of work, Wallace Arthur (2000, 2001, 2002a, 2002b, 2004, 2015) has shown that developmental bias can be quantified as a parameter in population genetic models. In these models, it serves as a direction-giving factor in evolution. The chance that evolution will proceed in a given direction depends on both (a) the probability that mutations will be generated in that direction and (b) the probability that such mutations will spread.

Consider the reversion of domesticated species to phenotypes resembling their "wild" types. At first glance, a purely selective explanation of this phenomenon seems sufficient. In domesticating a species, humans impose capricious selection pressures. The associated phenotypic changes are deleterious in the wild, and so on release revert to something like their original form. However, this is not the full story. Artificial selection frequently pushes traits to extremes, taking developmental systems as far as they will readily go in a given direction. When this occurs, the vast majority of mutations will experience developmental drive in the direction of reversion.

Because Arthur's work integrates developmental bias into population genetic models (a core part of mainstream, functionalist evolutionary theorizing), I have discussed it under the heading of using structuralist reasoning to deepen functionalist explanations. However, what Arthur really shows is that developmental bias and natural selection are both population genetic causes – they interact on a level playing field. Whether bias and selection are complementary (both favor the same phenotypes) or antagonistic (they push in opposed directions), their interactions do not involved the subordination of developmental bias to natural selection or vice versa.

3.3 Integration II: Using Functionalism to Deepen Structuralism

Explanatory integration of the sort just considered does not bring out the tension between evo-devo and mainstream theorizing, because it concerns ways in which the former can contribute to the latter's functionalist explanatory projects. While the *cis*-regulatory hypothesis did spark a small kerfuffle (Hoekstra and Coyne 2007; Carroll 2008; Stern and Orgogozo 2008; Craig 2009), it did not concern evo-devo's structuralism. Deeper tensions emerge when we turn to evo-devo's understanding of the sources of developmental bias.

Explanations in evo-devo frequently invoke deeply conserved developmental types to explain evolutionary trends well above the species level. Biases in the generation of variation are causally downstream of these types. But "type" is a fraught notion, and evo-devo has not escaped the fray. My aim is threefold: (1) to demystify evo-devo's invocation of developmental types by (2) illustrating another way that evo-devo integrates structuralist and functionalist explanatory strategies, thereby (3) demonstrating that evo-devo is an *essentially* hybrid structuralist-and-functionalist research program.

The modern synthesis, from which the current functionalist mainstream emerged, abhorred typological reasoning (Mayr 1959; Hull 1965a, 1965b; Sober 1980), though "typological thinking" has proven an elusive target (Chung 2003; Witteveen 2015, 2016, 2018). Modern synthetic theorizing was resolutely gradualist: small changes accumulate over time, primarily due to selection. Macroevolution (e.g., speciation, novelty) is just compounded microevolution (small changes within populations). Critics of synthetic theorizing frequently targeted its gradualism (e.g., Goldschmidt 1982 [1940]; Schindewolf 1993 [1950]). Ernst Mayr, concerned to understand the roots of such saltationist criticisms, found a diagnosis: the critics were in the thrall of typological thinking.

Mayr's (1959, 1960) basic idea of the connection between saltationism and typology ran as follows. Saltationists believe that selection's ability to accumulate small changes is limited. The location of this limit varies: for some, microevolution is restricted within species limits (and so impotent to explain speciation); for others, microevolution is restricted to tinkering with existing structures (and so impotent to explain novelties); other limits are also possible. Wherever the limit falls, however, only saltational change can cross it: small variants are constrained within the limit. Types were the source of these limits. Types existed over and above the actual variation present in a population. While saltationists allowed that types could change, this could only occur by processes radically distinct from those driving within-type microevolution. Thus Mayr (1960, 364) concluded that "all saltationists have been typologists, and most typologists have been saltationists of one sort or another."

The case of digit reduction in amphibians (Section 2.6) exemplifies "typological" structuralism in evo-devo. Alberch and Gale (1985) explain trends in digit reduction in frogs by treating them as modifications of a fixed type: the "generalized frog foot." This generalized foot renders some digit reduction phenotypes possible, others impossible. Although the generalized frog foot is both the product of past evolution and potentially subject to future evolution, Alberch and Gale treat it as a fixed source of biased variation in their explanation. What is the justification for doing so? Even if we historicize Unity of Type and trace types to inherited ancestral states, they can still seem mysterious. As Reeve and Sherman (1993, 19) put it, echoing Mayr's concerns, "ancestral species do not . . . mysteriously reach from the past to clutch the throats of their descendants."

The defender of developmental types thus bears a two-fold burden. First, they must explain *how* ancestrally inherited types "clutch the throats" of current lineages – that is, they must ontologically demystify them. Second, they must explain *why* ancestrally inherited types "clutch the throats" of current lineages.

Bias-generating features of organisms are themselves capable of evolving, so why do they persist? Both questions can be given compelling answers, but only if we recognize how evo-devo's invocation of types is integrated with functionalist reasoning. This will allow us to see how type-explanations are compatible with gradualism – while typologists are *often* saltationists, they are not *necessarily* so.

Having discussed the first issue at length elsewhere (Novick 2019; see also Lewens 2009), I will give only a brief overview here. In my earlier citation of Reeve and Sherman (1993, 19), I excised some crucial context. Here is the full quotation (emphasis added):

> Whatever is important about phylogenetic history will be recorded in the species' current environment and biological attributes. Ancestral species do not *otherwise* mysteriously reach from the past to clutch the throats of their descendants.

Reeve and Sherman do not reject the possibility of phylogenetic history affecting current evolution outright, so long as the right conditions are met. Specifically, only those aspects of that history that are "recorded" in the current traits of a species can shape that species' evolution. Only if this condition is not met are types mysterious.

This is all the defender of developmental types needs. The generalized frog foot influences the evolution of present-day anurans because it is a conserved feature of that lineage. Because it is shared by all anurans, it generates the same biases in lineages that are otherwise evolving independently. That is the sense in which it is "recorded" in the "current . . . biological attributes" of anurans. There is no deep ontological mystery here.

But this immediately raises the second problem: explaining why developmental types are conserved. After all, the generalized frog foot is not an eternal verity: it is an evolved feature of a particular lineage. It is not identical to the generalized salamander foot, with which it shares an ancestor – at least one of the two must have changed at the base of the anuran/urodele split. But this was a one-off event, and subsequent limb evolution has occurred within the constraints set by the respective types. Why? If the generalized frog foot is capable of changing, why – across all extant, independently evolving frog lineages – hasn't it? What accounts for the relative evolutionary stasis of developmental types (Wake, Roth, and Wake 1983)?

One possibility is that developmental types are conserved because they evolve by distinct and relatively rare processes. On this view, while evolution within types proceeds by the gradual accumulation of mutations with small effect sizes, the evolution of types involves different, likely saltational modes of evolution. Saltationist critics of the synthesis took this view, and some contemporary structuralists defend it as well (e.g., Newman 2016, 2021). I will return to it later (Section 4). Here, however, I hope to show that evo-devo's invocation of types can be justified entirely within a gradualist framework; it does not *require* saltationism.

A second possibility is to argue that the evolution of types is heavily *functionally* constrained. On this view, although types *can* change via gradualist processes, they do so relatively rarely, because such changes are rarely adaptive. Explanations of the deep conservation of types in evo-devo often take this second form. Before diving into the details, a word of caution. Not all bias-generating features of organisms are deeply conserved. Some biases really are weak and local. Moreover, of those that are deeply conserved, the phylogenetic depth of conservation varies (Peter and Davidson 2011). To successfully explain the relative evolutionary stasis of developmental types, one must be able to account for this variation.

This brings us back to the *cis*-regulatory hypothesis. Though I presented it in isolation in Section 3.2, it is actually only the tail end of a much more general picture of regulatory evolution. The *cis*-regulatory hypothesis claims that not all regions of the genome are created equal: there are systematic differences between mutations affecting protein-coding regions (which have highly pleiotropic effects) and mutations affecting regulatory regions (which have more modular effects). Because the latter have less pleiotropic effects, they are more likely to be selected. Regulatory DNA is thus expected to be more evolutionarily labile than protein-coding DNA.

True so far as it goes, but it only goes so far. Indeed, not all regions of the genome are created equal, but this cuts deeper than the *cis*-regulatory hypothesis, in isolation, recognizes: not even all *regulatory* regions of the genome are the same. Gene regulation in development involves structured, hierarchical regulatory networks (Peter and Davidson 2015; Figure 2). Subcircuits at different levels of this hierarchy serve systematically different functions (e.g., axial patterning, body part specification, and body part formation; Peter and Davidson 2011). Subcircuits that serve these different functions in turn have systematically different network topologies (Peter and Davidson 2015, chap. 6).

These differences in function and topology cause a third difference: subcircuits at different levels are evolutionarily labile to different degrees (Peter and Davidson 2015, chap. 7). This is the crucial difference for my purposes. What is it that explains this variation in evolutionary lability? It cannot be that variants affecting conserved regions of these networks are less likely to arise. Chemically speaking, it's all just DNA, all equally subject to replication errors and other sources of mutation. The constraints on network evolution are thus not structuralist.

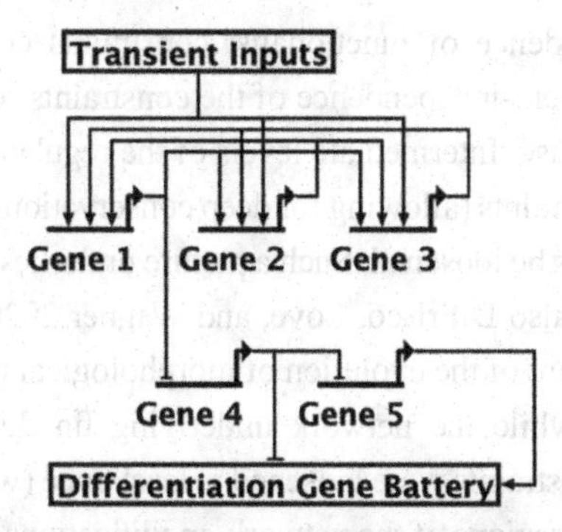

Figure 2 A hypothetical gene regulatory network, depicting a series of regulatory interactions (between genes 1 and 5) that convert a set of transient inputs into the (non-)expression of a differentiation gene battery (genes that define the differentiated state of the cell). Arrowheads indicate positive regulatory interactions; flat heads indicate repressive regulatory interactions. Created using BioTapestry (Longabaugh, Davidson, and Bolouri 2005, 2009)

Rather, subcircuits at different levels of the regulatory hierarchy are conserved to different degrees because they are subject to different strengths of *functional* constraint. At the deepest levels of conservation, the GRN "kernels" that have been conserved for hundreds of millions of years (Davidson and Erwin 2006), these functional constraints are incredibly strong. Genetic mutations that change the wiring of kernels are almost uniformly *nonviable* – either lethal or, if nonlethal, still severely deleterious (Peter and Davidson 2015, chap. 7). Moreover, this nonviability is *environment-independent*: there are few if any natural environments in which disruption of a regulatory subcircuit essential for heart development is viable.

This strong, environment-independent functionalist constraint on the evolution of kernels can then explain why early developmental processes are so deeply conserved (Raff 1996; Peter and Davidson 2015, chap. 7; Erwin 2020). Environment-independence matters because traits that are deeply conserved have to remain constant over time (even as environments change) and across different lineages (which may inhabit widely different environments). Such explanatory strategies, which rely on strong, environment-independent functional constraints, are "Cuvierian functionalist," in contrast to Darwinian functionalist explanations, which focus on the relationships between organisms and particular environments (Novick 2019).

Thus, while the *cis*-regulatory hypothesis predicts that mutations driving the evolution of development will largely occur in regulatory regions, the reality is more complicated: not all regulatory regions are equally modifiable.

Environment-independence of functionalist constraints comes in degrees: the almost total environment-independence of the constraints restricting kernel evolution is an extreme case. Intermediate levels of the regulatory hierarchy may be subject to strong constraints (allowing for deep conservation) that can nonetheless in some circumstances be loosened. Such a picture underlies, for instance, Günter Wagner's (2014; see also DiFrisco, Love, and Wagner 2020; DiFrisco, Wagner, and Love 2022) account of the evolution of morphological novelties (e.g., the fin to limb transition): while the network underlying fin development is highly constrained, these constraints were loosened at least once (when fins transformed into limbs). Terminal regions of the network, modifications to which are responsible for much of the minor morphological evolution we see, are even less tightly constrained (Davidson and Erwin 2006; Peter and Davidson 2011, 2015, chap. 7).

The upshot is a sliding scale of evolutionary lability: different regions of the regulatory network are labile to different degrees. To a first approximation, deeper regions of the network are less labile than shallower regions, though things are not quite so simple. Rather, particular functional subcircuits are the loci of deep conservation; not all regulatory linkages, however deep in the hierarchy, are part of such subcircuits, so there are highly labile linkages at all levels of the hierarchy. However, the most deeply conserved subcircuits do tend to be deep in the hierarchy, as they tend to have the most downstream dependences and thus the tightest functional constraints (Davidson and Erwin 2006; see also Schank and Wimsatt 1986).

We are now in position to tie together the various strands of this rather long and tortuous section. Evo-devo is structuralist insofar as it explains important aspects of the evolution of form in terms of the operation of generative processes that structure the space of possible form. Moreover, in these explanations, these generative processes are treated as fixed types, even though they are themselves evolved – even though they arose at some particular point in evolutionary history, and might change or disappear in the future. Our problem was to explain what justifies this.

Though it took a while to get there, the answer is in the end fairly simple: the features that evo-devo biologists identify as types are indeed evolved and subject to change, but they change relatively rarely, and so are conserved at deep taxonomic levels. Because of this, they can explain similarities in the space of possible form at taxonomic levels well above the species. This is so even if, when they do change, the process is a standard, gradualistic accumulation of mutations of small effect size. While the view that the evolution of developmental types occurs via different processes than evolution within types remains a live option (Newman 2016, 2021), it is not *necessary* to make sense of evo-devo's invocation of types.

Moreover, if we ask *why* these types are deeply conserved, the answer is functionalist: mutations that alter them are, for various reasons, unlikely to be selected. Though explanations that invoke developmental types do realize the structuralist explanatory strategy, they depend for their power on the phylogenetic distribution of those types. Explanations of that distribution realize the functionalist explanatory strategy: types are, to varying degrees, "locked in" for functionalist reasons. Thus, even the most notoriously "structuralist" explanations in evo-devo prove to be essentially integrative. Evo-devo's structuralism is inseparable from its functionalism. Viewing evo-devo as a purely or even predominantly structuralist research program obscures its true nature.

3.4 Integration III: Neutral Evolution Enters the Fray

Structuralist and functionalist strategies do not exhaust the evolutionary biologist's explanatory toolkit. In particular, the rise of mathematical population genetics (Provine 2001 [1971]) revealed the importance of a third explanatory strategy: neutralism (Kimura 1983; Ohta 1992). While this is not the place for a detailed study of neutralist explanations, their integration with structuralist and functionalist explanations deserves consideration. I will consider two examples: (1) constructive neutral evolution, and (2) Andreas Wagner's (2011) model of evolutionary innovation.

Natural selection is sometimes claimed to be the only evolutionary cause capable of reliably increasing organismal complexity (Dawkins 1986, chap. 11). While any individual neutral change must of course proceed in *some* direction – including, potentially, that of increased complexity – it is hard to see how it could consistently accumulate changes in a particular direction. Each neutrally fixed mutation is (by definition) fixed by chance and thus reversible: the next neutrally fixed mutation might equally undo it.

Though this argument is tempting, it does not work. The basic issue is that, while each individual change fixed by drift is reversible *at the time of fixation*, later neutral changes can render earlier ones irreversible. Under the right conditions, this can allow a sequence of exclusively neutrally fixed mutations to preferentially increase complexity, a process known as constructive neutral evolution (CNE). Originally developed in the context of molecular and cellular evolution (Stoltzfus 1999, 2012; Gray et al. 2010; Lukeš et al. 2011; see Speijer 2011 for a critique), CNE has since been applied more broadly (Brunet and Doolittle 2018; Wideman et al. 2019; Muñoz-Gómez et al. 2021; Brunet 2022).

CNE can be illustrated with a simple example (drawn from Doolittle 2012). Imagine a functional protein complex with eight identical subunits, all encoded by a single gene. Let this gene be duplicated (call the two versions of the gene

"A" and "B"), such that some subunits are generated from A and some from B. Over time, A and B will diverge by the accumulation of neutral mutations. This divergence will be constrained by the need for the full complex to maintain its function: neutral changes in A will alter which changes in B are neutral, and vice versa. This allows later changes to "lock in" earlier changes. While a change to A may have been neutral (and so reversible) when it was fixed, subsequent change to B can make reversion deleterious. By such means, the overall complexity of the protein complex irreversibly increases, even though each individual complexity-increasing mutation was fixed neutrally.

Already we can see how constructive neutral evolution essentially integrates functionalist and neutralist explanatory strategies: neutral evolution explains the initial fixation of complexity-increasing mutations, while functional constraints lock that complexity in place. These functional constraints render (most) complexity-decreasing mutations deleterious, leaving drift no options but to maintain or increase complexity. Even drift can be directional, if it only has one direction to go.

Can structuralist considerations be integrated as well? I think they can. One of Stoltzfus' (1999) initial applications of CNE was to gene duplication, which has played an extensive role in the evolution of regulatory DNA in metazoans, in particular of the *hox* genes (Duboule 2007). This suggests the tantalizing hypothesis that CNE has played an important role in the evolution of metazoan development. I must stress that the account below is speculative: it shows how the three explanatory strategies *might* be integrated (cf. Prince and Pickett 2002), but I make no claim that the resulting explanation is correct.

Though all animals share a conserved core of regulatory genes (of which *hox* genes are an important subset), different taxa vary wildly in the number of *hox* genes they possess (Lemons and McGinnis 2006; Duboule 2007). In numerous lineages, either individual *hox* genes or entire hox gene clusters have undergone one or more rounds of duplication, followed by (in some cases) subfunctionalization. This process is thought to be highly important for animal evolution (e.g., Stellwag 1999; Soshnikova et al. 2013). Though the duplication events lie too deep in the past to ever know if they were neutral, it is at least plausible that they are a case of constructive neutral evolution (cf. Greer et al. 2000).

The expansion of *hox* gene clusters has two important effects. First, duplication can relax some of the functional constraints on gene network evolution, because only one of the copies need preserve the ancestral function. This can then allow the evolution of conserved types (Section 3.3). Second, duplication can increase the complexity (and potentially depth) of the gene network hierarchy. This can generate both novel functional constraints (on deeper regions of the hierarchy) and novel developmental biases.

Thus, on the plausible assumption that constructive neutral evolution played a role in *hox* gene cluster expansion and diversification, there is a complex interplay between neutral evolution, functional constraints, and structural biases. On this view, neither a purely functionalist, nor a purely structuralist, nor a purely neutralist account of the evolution of animal form will do – only an integrated explanation.

Andreas Wagner's (2011) work on evolutionary innovation reveals another way of integrating neutralism with functionalist and structuralist strategies. The basic idea is that neutral evolution enables the "silent" exploration of possibility space. Imagine a long chain of metabolic chemical reactions that starts with some input molecule (gleaned from a food source) and ends with some output molecule (usable by the organism). Assume that, if the organism cannot produce the output from the input, the organism dies.

Between the input and output, there is a long chain of metabolic reactions that are functionally unconstrained (Wagner makes the idealizing assumption that only the input-output relation is visible to selection). So long as the chain yields the required output, any modifications to it are neutral, allowing for the radical revision of the chain's internal structure by neutral evolution. The functionally constrained phenotype remains constant, but the lineage wanders a subterranean neutral path.

A neutral walk of this kind can alter the *possible* directions of phenotypic evolution. As the internal sequence of reactions changes, different outputs become mutationally accessible. In this way, a "metabolic network … can gain access to a virtually inexhaustible reservoir of new phenotypes" (Wagner 2011, 29). Neutral evolution thus (a) modifies the structure of the space of possible forms, i.e. what phenotypes are mutationally accessible, and so (b) changes the possible directions of adaptive evolutionary change, enabling some innovations and forbidding others.

Thus we have another form of integrative explanation. We begin with functional constraints that require that a particular phenotype (input-output relation) be held fixed. This generates a set of neutral evolutionary paths (whatever mutations change the internal sequence of reactions without changing the ultimate input-output relation) that allow a lineage to silently explore a space of possibilities (altering which regions of this space are accessible), which ultimately determine which innovations (changes to input-output relation) are even potentially available for natural selection to act on. Wagner's explanation of innovation moves from functionalist to neutralist to structuralist reasoning, finally returning to functionalist concerns.

3.5 Interim Conclusion II

Much of the philosophical interest in evo-devo has focused on its alleged status as a structuralist challenger to a functionalist mainstream. This interest is not altogether misplaced: evo-devo does make extensive use of the structuralist explanatory strategy, and this does generate points of tension with mainstream theorizing. However, it is misleading to characterize evo-devo *as a whole* as "structuralist." Evo-devo is better understood as omnivorous and integrative: omnivorous because it makes use of the full range of explanatory strategies available within evolutionary biology; integrative because many of its most compelling explanatory successes rely, not on applying these strategies in isolation, but rather on binding them together into complex explanatory chains.

4 Integration without Subordination

4.1 Subordination Arguments

Analyzing structuralism and functionalism in terms of explanatory strategies renders the application of "structuralist" and "functionalist" to individual scientists somewhat nebulous. If we recognize that the structure of evolutionary theory (and the fact of anisotropic variation) requires all biologists to accept the legitimacy of both strategies, the difference between "structuralists" and "functionalists" seems to become both contextual and a matter of degree (Section 2.6). Recognizing further that evolutionary theory gains in explanatory power when the strategies are integrated (Section 3), we may also wonder whether anyone *ought* to be a "structuralist" or a "functionalist," even in a weaker sense.

But this is too fast. There does seem to be *some* sense in which Owen, whose archetype theory severely limited the domain in which functionalist explanations are legitimate, is plainly a structuralist. Conversely, Darwin, who argued that Unity of Type is just the byproduct of Conditions of Existence plus time, seems equally plainly a functionalist. Despite acknowledging a role for both strategies, Owen and Darwin both treat one as more fundamental than the other. We might thus identify biologists as functionalists and structuralists on the basis of their attempting to subordinate one explanatory strategy to the other – on the basis of their offering what I shall call *subordination arguments*.

Such arguments are common in the history of biology, and so merit close examination. In what follows, I develop a conceptual framework for understanding such arguments (this section), then provide a general assessment of their force (Section 4.2). After this, I return to the question of what it might mean to call an individual "structuralist" or "functionalist" (Section 4.3). As I discuss them, subordination arguments concern empirical claims about where structuralist and functionalist explanations apply and how they relate. They do not concern value judgments about what it is most important to explain (cf. Boucher 2015). I defer discussion of the latter to the conclusion (Section 5.1).

Though all subordination arguments ultimately conclude that one explanatory strategy is subordinate to the other, they vary widely in their details. Here is a nonexhaustive array of examples:

1. **Functionalist subordination argument 1.** In *De Anima* II.4, Aristotle raises two problems for Empedocles' explanation of plant growth in terms of the natural tendencies of earth (to move downward) and fire (to move upward). First, it misconceives the *explanandum*: the very *identity* of parts can only be understood by understanding their function. Functionally considered, roots are the *top* of a plant, not the bottom; thus Empedocles' "conception of up and down is wrong" (Irwin and Fine 1996, 88). Second, Empedocles cannot explain the *integrity* of plants, given that fire and earth move "in contrary directions" (Irwin and Fine 1996, 88). Without some *limit* to these natural tendencies, the earth and fire will separate. For Aristotle, the source of these limits is the organism's soul, understood as the its primary functional capacities (Lennox 2021). Organic structure is to be explained by these capacities. Though not *irrelevant* to explaining plant growth, material causes are *subordinate* to functional causes.
2. **Functionalist subordination argument 2.** Though he bitterly opposed Geoffroy's attempts to find structural correspondences between *embranchements*, Cuvier allowed that each *embranchement* was characterized by a structural plan shared by each species within it. However, this plan was a *functionally integrated* unity. Cuvier deemed forms intermediate between *embranchements* to be impossible, not on structural grounds, but because they are not *viable* (Novick 2019). Insofar as Unity of Type captures a real phenomenon, its explanation is functional.
3. **Functionalist subordination argument 3.** Darwin (1964 [1859], 206) argued that Conditions of Existence is a "higher law" than Unity of Type, on the grounds that Unity of Type is simply what results when Conditions of Existence operate across time, rather than all at once (Section 2.2). This allows for structural correspondences to be independent of *present* function, but they are not independent of function altogether. Rather, parts are shaped by both present *and past* functions. As with Cuvier, insofar as Unity of Type is recognized as a real phenomenon, it receives a functional explanation.
4. **Functionalist subordination argument 4.** The architects of the modern synthesis argued that natural selection is the sole direction-giving factor in evolution (Mayr and Provine 1981). While allowing the existence of minor

nonselective sources of direction (e.g., correlated growth; Section 2.4), they insisted on the primacy of natural selection. This had both a negative and positive aspect. Negatively, the view excluded alternative theories positing nonselective sources of direction (e.g., Goldschmidt 1982 [1940]; Schindewolf 1993 [1950]). Positively, the view asserted the *creativity* of natural selection: selection directs evolutionary change by "creating the variation that it subsequently acts on" (Beatty 2016, 2019), allowing features "that are otherwise very unlikely to arise via the immediate sources of variation [to] become much more likely to arise" (Godfrey-Smith 2009, 50). Combined with the Haldane-Fisher argument for the unimportance of mutation biases (see Yampolsky and Stoltzfus 2001), this established selection as playing the dominant role in determining evolutionary direction, with structuralist explanations left on the sidelines.

5. **Structuralist subordination argument 1.** Richard Owen (2007 [1849]) gave new life to structuralist thought in England, in the wake of the Cuvier-Geoffroy dispute, with his notion of the vertebrate archetype. The archetype was the structural core of an ideally simple vertebrate, consisting of a number of repeated segments (serial homologs). The forms of all actual vertebrates were modifications of this structural core, containing the same parts in the same arrangement (special homologs). These modifications could be explained functionally, but the underlying structure itself could not. Owen thus relegated functional explanation to secondary status: it could explain modifications within the confines of the archetype, but not the archetype itself.
6. **Structuralist subordination argument 2.** Richard Goldschmidt (1982 [1940]) defended a saltationist, structuralist version of evolutionary theory. He allowed that the accumulation of micromutations by natural selection could explain evolution below the species level (microevolution). However, the evolution of new species and new morphological structures (macroevolution) occurs by a different process: a radical reorganization of the developmental system that generates a novel form in a single step. Selection can pass a simple yes/no verdict on these "hopeful monsters," but the direction of the change is determined by the potentials of the developmental system itself. Goldschmidt thus restricted the role of functionalist reasoning to a narrower (and less important) domain than structuralist reasoning.
7. **Structuralist subordination argument 3.** In his wonderfully titled "The Logic of Monsters," Pere Alberch (1989) sought to explain the "discreteness and order" of natural diversity, i.e., the fact that morphospace is occupied in a clumpy rather than in a smeared fashion. He considers and rejects the functionalist (he calls it "externalist") explanation that this clumpy

distribution follows the topography of the adaptive landscape. Instead, he argues that "generative rules" underlying development determine which regions of morphospace are even potentially occupiable. Selection is limited to determining which regions of occupiable morphospace are in fact occupied.

8. **Structuralist subordination argument 4.** Biologists working on "dynamic patterning modules" challenge the view that the early evolution of form in multicellular (Newman and Bhat 2008, 2009; Hernández-Hernández et al. 2012; Benítez et al. 2018) and aggregative organisms (Arias Del Angel et al. 2020; Guzmán-Herrera et al. 2021) involved selection favoring regulatory mutations of small effect size. Instead, they argue, basic body plans are explicable in terms of the action of ancient genes whose products affect the physical properties of cells (e.g., cell–cell adhesion). These processes generate a basic space of possible forms that arise without reference to any particular function. Organisms can then seek out niches in which these forms are beneficial, allowing natural selection to play the secondary role of stabilizing and canalizing them (Niklas and Newman 2013).

Henceforth, I will use the shorthand FSA1-FSA4 for the functionalist arguments, and SSA1-SS4 for the structuralist arguments (Table 2).

These examples bring to light some general features of subordination arguments. It is essential to such arguments that both explanatory strategies are allowed to play a legitimate role, though one is, of course, relegated to a subordinate role. In this sense, subordination arguments are at least minimally integrative: they allow both strategies to peacefully coexist – so long as the subordinate one knows its place. The subordinate strategy might be legitimate only within narrow confines (e.g., a functionalist who accepts only weak and local developmental biases, or a structuralist who allows cumulative selection to explain microevolution only), but it is not *totally* excluded.

Beyond those core similarities, however, the arguments vary widely. Most obviously, they differ in which strategy they treat as subordinate. No more need be said about this. Two other differences deserve consideration: these concern, not *which* strategy is subordinate, but rather (a) the *domain* in which it is subordinate and (b) *how* it is subordinate. These differences point to interesting general features of both FSAs and SSAs – they are not random, but rather explicable in terms of the kinds of arguments FSAs and SSAs are.

First, some subordination arguments are *essentially* restricted to a particular domain, while others are at least *potentially* universal. SSA2 (Goldschmidt) and SSA3 (Alberch), which reference features of multicellular development, are essentially restricted, since not all life is multicellular. SSA4 (Newman and Bhat)

Table 2 Structuralist and functionalist subordination arguments. Further explanation in text

	Basis of subordination	Select proponents	Essentially restricted?	Principled kinds or relative significance?
FSA1	Final causes explain the integrity of the organism and the limits to material causes	Aristotle (Irwin and Fine 1996; Lennox 2021)	No	Significance
FSA2	Functional integration rules out intermediate forms as nonviable; this explains Unity of Type	Cuvier (Novick 2019); Davidson (Davidson and Erwin 2006)	No, but has restricted variants	Significance
FSA3	Unity of Type is reducible to the operation of Conditions of Existence over time	Darwin (1964 [1859]); Beatty (2016)	No	Significance
FSA4	Selection is the sole direction-giving factor in evolution, and is *creative* of the variation it acts on	Modern synthesis (Godfrey-Smith 2009; Beatty 2019)	No	Significance
SSA1	Archetype constrains fundamental form; functional modifications are secondary and limited	Owen (1848, 2007 [1849])	No	Kinds
SSA2	Speciation occurs by macromutation, with selection limited to yes/no verdicts	Goldschmidt (1982 [1940]); Schindewolf (1993 [1950])	Yes	Kinds

SSA3	Generative rules of development determine *occupiable* regions of morphospace; selection determines actual *occupation*	Alberch (1989)	Yes	Kinds
SSA4	Physical forces dominated early metazoan evolution; selective fine-tuning via micromutation became possible only later	Thompson (1992 [1942]); Newman and Bhat (2008, 2009)	Yes, but has unrestricted variants	Kinds

also requires multicellularity or aggregativity, and so is essentially restricted as stated. However, it might be seen as a special case of a more general subordination argument, one based on the fact that physical forces shape the possible forms for all life, though different specific forces dominate in different domains (Thompson 1992 [1942] may provide such an argument). That argument would be potentially universal. The other five arguments – notably, all the FSAs – are all potentially universal.

What accounts for the fact that the FSAs are all potentially universal, while the SSAs are mixed? Does this reflect a real difference between FSAs and SSAs, or is it an artifact of my chosen examples (after all, my sample is neither large nor random)? Recall that structuralist explanations cite generative processes that structure the space of possible form. Some causes that affect this structure are universal (e.g., physical forces as such); others necessarily affect only some lifeforms (e.g., multicellular development). That is why SSAs are mixed. By contrast, all kinds of life must be able to survive and reproduce in some environment – all, that is, must function. If a principled reason can be found to think that functionalist explanations are dominant and structuralist explanations are subordinate, it is likely to apply universally, at least potentially. However, restricted FSAs are possible. Consider FSA2 (Cuvier), which is based on the presence of tight functional integration between the parts of organisms. Cuvier intended it to apply universally, but we now know better: some aspects of organisms are tightly integrated, but not all. Any contemporary version of FSA2, therefore, would be restricted (Novick 2019).

The second difference concerns the manner in which one explanatory strategy is rendered subordinate to the other. Some subordination arguments draw a principled division between the kinds of cases where each strategy dominates, while others concern only the relative significance of each cause.

All four SSAs take the first form. Structuralist explanations invoke generative processes that structure the space of possible form. This allows biologists to distinguish the task of explaining why the space has this structure from the task of explaining how actual forms are distributed within it. Insofar as structuralist strategies dominate explanations of the former, an SSA is the natural result. For example, SSA3 (Goldschmidt) distinguishes macroevolution from microevolution, arguing that the former *cannot* occur by the accumulation of small mutations. Though functionalist explanations are given a territory (microevolution) in which they are legitimate, the evolution of major aspects of form falls within structuralist territory (macroevolution). I say "territory" rather than "domain" to keep this axis separate from the last: domains concern kinds of organisms (e.g., microbes vs. macrobes); territories concern explanatory tasks within domains (e.g., macrobial macroevolution vs. macrobial microevolution).

By contrast, all four FSAs concern the relative significance of structuralist and functionalist explanatory strategies, without drawing any principled distinctions between kinds of cases. FSA4, for instance, concerning the "creativity" of natural selection, is meant in part to justify the view that macroevolution is just compounded microevolution. The idea is that, by continually opening up new variational possibilities to explore, natural selection can eventually produce even large-scale changes in form. The same process (natural selection accumulating micromutations) dominates everywhere (a similar idea animates Darwin's FSA3). That is why functionalists of this sort argue that developmental biases are weak and local (Alberch and Gale 1985): they share a territory with selection, and so must be weak enough to be subordinated within that territory. For FSA1 (Aristotle) and FSA2 (Cuvier), the rejection of a principled distinction turns on the fact that *every* aspect of an organism is to be understood in terms of its function within the whole. It is not that structuralist explanations (e.g., in terms of the natural tendencies of earth) have no role, but where apt they are *always* subordinated to functionalist explanations.

4.2 Integration without Subordination

So much for the nature of subordination arguments – are any of them compelling? Do we have good reason to think that one or the other explanatory strategy is dominant, and the other subordinate? More specifically, I want to ask (a) what conditions must be met for a subordination argument to succeed, and (b) what the prospects are for meeting those conditions. My focus is on the *general* prospects of subordination arguments, rather than the status of any one in particular. Because structure/function disputes today take place in an evolutionary context, I will set aside Aristotle's FSA1 and Owen's SSA1. Cuvier's FSA2 has evolutionary descendants that are relevant here.

I contend that *all* universal subordination arguments fail: we do not have good reason to think either explanatory strategy is *generally* dominant. Subordination arguments can be compelling, but only within restricted domains. In defending this view, I will rely on quite general features of evolutionary theory. Because of their generality, they furnish good (if defeasible) reason to expect that any future subordination arguments will fail unless they are suitably modest. The upshot is a view of the relationship between the two explanatory strategies that stresses their potential for integration as equals.

Each of the universal subordination arguments (FSA2-4, and the generalized version of SSA4) begins from an important truth about how evolution operates.

However, none of these truths are sufficient to establish the dominance of one strategy over the other. Consider Darwin's FSA3 first, which argues that Unity of Type is the result of the inheritance of past adaptations and is therefore a byproduct of Conditions of Existence. The core truth here is that deep structural similarities are themselves the product of past evolution (potentially adaptive, but possibly neutral, as we now know). This is only half the picture, however. Darwin conceives of Unity of Type merely as inherited similarities, without recognizing how these conserved features bias variation. Unity of Type is not just a byproduct of Conditions of Existence, but also shapes its operation – the interaction between them is bidirectional. The mere fact that bias-generating features have themselves evolved does not show that functionalist explanatory strategies are fundamental, *even if those features evolved by selection*.

A similar issue besets Cuvier's FS2 (in its evolutionary variants) and FS4 (on the creativity of selection). Cuvier's point that tight functional integration renders intermediate forms nonviable plays an important role in contemporary evo-devo (Novick 2019). Insofar as organisms show such tight integration (which is not nearly so far as Cuvier himself thought), they experience strong functional constraints on evolution, and this can help explain why certain bias-generating features are deeply conserved (Section 3.3). Does this show that structuralist explanations citing those bias-generating features are subordinate to functionalist explanations of their deep conservation? It does not, because those functionalist constraints are in turn given a structuralist explanation. Certain regions of the genome are difficult to modify *because of how they are employed during development* – because development channels mutations affecting those regions toward nonviable phenotypes. This is a case, not of subordination, but of thoroughgoing integration.

FSA4, which concerns the creativity of selection, similarly fails. Like the others, the basic idea is correct, even profound, but it fails to establish the dominance of functionalist strategies. FSA4 challenges the view that, since selection can only choose among the variants presented to it, it is limited to a filtering role. Against this, FSA4 notes that, since the features of organisms that generate variation are themselves evolved, natural selection shapes the variation that is available to it. It is not merely a filter. As a defense of the creativity of selection, FSA4 is sound. However, this only establishes that structuralist and functionalist explanations are entangled, not that structuralist explanations are subordinate to functionalist explanations. To get *that* consequence, FSA4 must be conjoined to an argument that structuralist biases are weak and local, and so dominated by selection. This was, historically, achieved by the Haldane-Fisher argument that mutation is only a weak evolutionary force, but this argument rests on a reification of simplifying assumptions in population genetic models, which are designed in such a way as to relegate the generation of new mutations to the sidelines (Stoltzfus manuscript; cf. Yampolsky and Stoltzfus 2001). When the generation of mutations is properly considered, the conclusion that biases must be weak and local collapses.

So much for the universal FSAs. What about the generalized form of SSA4? This argument contends that physical forces shaping the generation of variation are a universal feature of evolution, although the most relevant forces and associated generative processes will depend on particular features of organism-environment relations: a pond-skater deals much more with surface tension, and much less with gravity, than a human (Thompson 1992 [1942], 67). This can be conjoined to arguments about organisms' agency (Lewontin 1983; Odling-Smee, Laland, and Feldman 1996): the forms produced by these generative processes may be stabilized precisely because organisms either choose or create niches in which those forms are beneficial (Newman 2022).

Though powerful, there are two reasons why this argument fails to establish the *universal* dominance of structuralist explanations. First, in allowing that the evolutionary relevance of physical forces is environment-relative, it leaves space for the creativity of selection: selection will shape which forces matter to a particular form of life. This is just the inverse FSA4's core flaw. Acknowledging organisms' agency in choosing their environment (including both selective pressures and physical forces) only amplifies the point. Niche construction explanations are functionalist: they invoke organisms' responsiveness to their own needs as *explanans* (see Brown 2022, 2n3 for an alternate view). The creativity of niche construction, like the creativity of selection, shows how structuralist and functionalist explanations are entangled. Second, though physical forces are universal, this says nothing about how wide or narrow a range of variation they allow. Both the nature and the strength of physical forces' effects on the structure of the space of possible form will vary with organism-environment relations.

Though I've diagnosed individual failings for each of these universal subordination arguments, there is a more general reason why all of them fail – why *any* universal subordination argument is likely to fail. This reason turns on two basic features of evolutionary theory, acceptable to structuralists and functionalists alike. First, at a global level, the structure of evolutionary theory is *integrative*. Second, evolutionary theory treats the course of evolutionary history as *contingent*.

For evolution to occur at all, variants must be generated, and those variants must spread through a population. Structuralist explanations invoke processes that bias the generation of variation; functionalist explanations invoke functional causes targeting either generation or spread. Neither role is *inherently* privileged;

rather, both are equally essential. It might seem like the generation of variation is fundamental, since only actually generated variants can spread, but this appearance is spurious. The process is iterative, and the spread of variants at one time influence which variants are subsequently generated (FSA4).

This point about the integrative structure of evolutionary theory establishes an abstract equality of the two strategies. However, once we consider concrete details, one strategy might prove more fundamental than the other. This brings us to the second feature of evolutionary theory: contingency (Beatty 1995; McConwell 2019). All life on earth is a contingent product of evolutionary processes. The contingent results of evolution themselves shape evolutionary processes: *evolution itself evolves*. Lateral gene transfer is far more common in prokaryotes than eukaryotes (Andersson 2005). The evolution of new hierarchical levels brings multi-level selection dynamics into play (Okasha 2006). Multicellular developers evolve subject to developmental biases; unicellular lifeforms do not. *There is no one way that evolution occurs, common to all life.* As these differences may affect the relative importance of structuralist or functionalist explanatory strategies, there is no reason to expect either strategy to be universally dominant.

The challenge evolutionary contingency poses to universal subordination arguments is subtle. The mere fact of contingency is not sufficient on its own to establish the failure of universal subordination arguments. After all, universal subordination arguments draw on those features of the evolutionary process that *are* universal. All evolutionary change is subject to the effects of physical forces (SSA4); all evolutionary change is iterative (FSA4); and so on. If any of these universal features of evolution were sufficient to establish that one strategy subordinates the other, contingency would be irrelevant.

However, the genuinely universal features of evolution are *too weak* to ground subordination arguments. Yes, physical forces are universal, but if they are highly permissive, they will have only weak effects, palatable to the functionalist, on the structure of the space of possible forms. Yes, evolution's iterativity allows selection to be creative, but if variation is sufficiently constrained, selection's creativity will be highly limited. To make a successful case for subordination, these general points must be conjoined to further claims, claims that concern precisely those features of organisms that *are* contingent and so are not universally distributed. Thus, evolutionary contingency precludes the universal dominance of either structuralist or functionalist explanations.

Matters are different for restricted subordination arguments. We have just seen that universal subordination arguments cannot succeed on their own, but rather must be conjoined to some further claim. Since these further claims concern contingent features of evolution, they will be restricted to particular domains. To augment a universal subordination argument with a domain-restricted claim is just to generate a restricted subordination argument. Under the right conditions, such arguments can succeed. Whether those conditions are met – whether the relevant domain-restricted claims are true – is an empirical matter. My aim below is simply to identify their success conditions. These differ between functionalist and structuralist arguments; I take them up in turn.

Restricted FSAs aim to show that, within a given domain, functionalist explanations have greater significance than structuralist explanations (Section 4.1). They do this by showing that, within that domain, biases affecting the generation of variation are both *weak* and *local*. The less that variation is channeled in a narrow range of directions, the weaker the bias. The narrower the taxonomic range affected, the more local the bias. The weaker and more local the bias, the less role it can play in explaining long-term, large-scale evolutionary change. While such biases will play a role in explaining particular changes, their cumulative effect will be haphazard and nondirectional. This leaves selection as the primary determinant of the direction of evolution. The problem with FSA4, a universal subordination argument, was that it merely showed the entanglement of structuralist and functionalist explanations. When conjoined to a domain-restricted claim that biases are weak and local, however, a restricted subordination argument is generated. So long as the domain-restricted claim is true, the restricted argument can succeed.

As for restricted SSAs, they must meet two conditions to be successful. First, they must establish the existence of a principled distinction between two kinds of cases, one of which must be more fundamental than the other. By "principled distinction," I mean that the two kinds of cases must be distinguished on the basis of differences in the causal processes responsible for them. Second, they must show that structuralist explanations dominate within the more fundamental kind of case. Both conditions must be met: merely establishing a principled division between cases does not yield a structuralist subordination argument by itself. Peter and Davidson (2015, chap. 7), for instance, draw a principled distinction between how different levels of the GRN hierarchy evolve, but offer functionalist explanations of evolutionary patterns at all levels (Section 3.3).

Work on dynamical patterning modules illustrates the role of both success conditions. This work rests on a distinction between (1) the *generation* of basic multicellular or aggregative morphologies and (2) the *stabilization* of these morphologies. This distinction is principled. The generation of basic morphologies is explained by the effects of dynamical patterning modules. A dynamical patterning module is "a set of molecules produced in a cluster of cells, along with one or more physical effects mobilized by these molecules so as to generate an aspect or alteration in the cluster's form or pattern" (Newman and Bhat 2008, 2).

By contrast, stabilization is explained by (a) organisms actively seeking niches in which the generated forms are beneficial and (b) selection and drift accumulating mutations of small effect size. Generation is more fundamental than stabilization and is dominated by structuralist explanations. Finally, the argument is restricted to those taxonomic domains in which either multicellularity or aggregativity has evolved. If sound, this argument meets both success criteria for restricted SSAs.

In sum: in light of evolutionary contingency, we should expect *all* universal subordination arguments to fail. Restricted subordination arguments may succeed, provided they satisfy the relevant success conditions. However, this will not yield a version of evolutionary theory that is either structuralist or functionalist *on the whole*. Rather, structuralist explanations will dominate for some phenomena in some domains, functionalist explanations for other phenomena in other domains, and integrative explanations still elsewhere. Considered as a whole, evolutionary theory can only be understood as integrative.

4.3 Structuralists and Functionalists

This section began in perplexity. My discussion of explanatory strategies (Section 2–3) left it rather nebulous what it might mean to call an individual biologist a "functionalist" or a "structuralist." Applying these labels on the basis of the mere *use* of one or the other explanatory strategy renders them highly contextual: whether a given biologist is a structuralist or functionalist will depend on who they are arguing against at any given time. This is illuminating in some cases. For instance, it helps us see that Günter Wagner's (2014) account of evolutionary novelty, commonly regarded as structuralist, is really integrative (Section 3.3). It seems structuralist because he most prominently takes functionalist alternatives as his foil, but would seem functionalist if compared to the views of someone like Stuart Newman (e.g., Stewart, Bhat, and Newman 2017).

However, some biologists do seem to have a deeper commitment to structuralism or functionalism; in such cases, it is unsatisfying to let the matter rest with a merely contextual designation. For some of these biologists, the deeper commitment may be to a stance; I consider such cases later (Section 5.1). Here, my focus remains on explanatory strategies. I consider two forms that a commitment deeper than the mere use of these strategies might take. The first is a commitment about how the strategies *relate* – that is, commitment to a subordination argument. The second is a commitment about how the proper application of the strategies is to be *discovered* – that is, commitment to methodology.

Universal subordination arguments plainly offer one way to identify individuals as either structuralist or functionalist: a structuralist (functionalist) is just someone who offers a universal structuralist (functionalist) subordination argument. This attributes a shared *kind* of substantive commitment to all structuralists (functionalists), while allowing that they need not share any particular commitments. This is compatible with, but does not require, their adopting a structuralist (functionalist) stance (Boucher 2015). I expect, by contrast, that any individual to whom such a stance is reasonably attributable is going to endorse *some* subordination argument, though possibly only a restricted one.

As a tool for making sense of historical (and some contemporary) figures, this manner of identifying structuralists and functionalists is useful. Many of the paradigm functionalists (Aristotle, Darwin, the synthesis architects) and structuralists (Owen, Thompson) can be found making universal subordination arguments. If my arguments in Section 4.2 are correct, however, nobody *ought* to accept universal subordination arguments, and thus nobody ought to be a structuralist or functionalist in this way.

What about restricted subordination arguments? While we might identify individuals who endorse such arguments as structuralists or functionalists, such identifications will be contextual. On this approach, a structuralist (functionalist) is someone who offers a restricted SSA (FSA) for a particular domain. Since offering such arguments in one domain does not require offering them in all domains, the same individual could be a "structuralist" in one context and a "functionalist" in another. There is nothing intrinsically wrong with this, though applying the labels in this way perhaps risks suggesting a more thoroughgoing commitment than the biologists in question actually hold.

Aside from subordination arguments, we might identify structuralists and functionalists on the basis of methodological commitments. There is already a large literature on methodological adaptationism (Godfrey-Smith 2001; Lewens 2008). The methodological adaptationist argues that evolutionary biologists should focus on *directly* investigating adaptive hypotheses. Given a trait of interest, the methodological adaptationist *assumes* that variation was roughly isotropic and that selection could therefore favor the optimal form. Proponents argue that, by using this assumption to develop testable hypotheses, biologists will discover more than just the role played by selection in evolution. They will also discover deviations from optimality (if there are any), allowing them to recognize constraints that might otherwise escape notice. In my terminology, methodological adaptationism is the prescription that one should default to offering functionalist explanations for all contrastive why-questions, not because those explanations are most likely to be true, but because doing so is a good heuristic for discovering which functionalist explanations are correct and which are not.

Equally possible, though less often discussed, is methodological structuralism (Brown 2022). The methodological structuralist identifies two limitations of methodological adaptationism. First, given sufficient ingenuity, potential adaptive explanations never run out (Gould and Lewontin 1979; see also Lloyd 2015). The promised revelation of constraints may never arrive. Even if it does, there is a second limitation. While methodological adaptationism *can* identify constraints, these are constraints on *function*, not constraints on *form* (Amundson 1994). It is the latter that feature in structuralist explanations.

To discover constraints on form, an alternative method is needed. The methodological structuralist notes that there is an important sense in which structure is prior to function: selection can only act on the variation provided to it. Variation must be generated before it can spread. Thus, before making assumptions about the role of selection in the evolution of a trait, it is important to understand how that trait can vary. Not only will this reveal constraints on form, it will also provide information about the range of variation available to selection. Methodological structuralism can thus help us to discover which why-questions will plausibly require functionalist answers.

In their study of amphibian digit reduction (Section 2.6), Alberch and Gale (1985, 18–19) used this approach to reveal roles for *both* structuralist *and* functionalist explanations:

> However, there are some differences between expectations from ontogeny and the morphologies observed in nature that do imply functional (selective) constraints on design. [...] This highlighting of dissociations from the expected patterns is one of the important insights of the developmentalist approach. The problem for a functionally oriented evolutionist is, now, not to explain the repeated reduction of the first toe in anurans but rather the persistence of the fourth phalange in the fourth toe.

By investigating the variation generated by the anuran and urodele developmental systems, Alberch and Gale determined that a particular why-question probably required a functionalist answer.

Gould and Lewontin's (1979) critique of adaptationism also advocates a form of methodological structuralism. They recommend *starting* one's analysis with careful attention to the problems of trait individuation before offering any functionalist hypotheses. These are structuralist problems, concerning what kinds of variations are possible, given a particular developmental system. While the *explanations* Gould and Lewontin offer are predominantly functionalist, they are improved by careful attention to structuralist concerns. At a methodological level, their recommendations are genuinely structuralist.

Based on the foregoing, methodological structuralism does appear to have a significant advantage over methodological adaptationism. Methodological adaptationism is limited in its ability to reveal constraints on form. Constraints on function may be evidence of constraints on form, but they are not the same. As a result, it is less useful for identifying where structuralist explanations are required than its proponents hoped. By contrast, methodological structuralism can identify why-questions requiring either type of answer. While it would be premature to recommend a thoroughgoing methodological structuralism on this basis, the issue deserves attention comparable to that lavished on its adaptationist counterpart.

So, what makes an individual a "structuralist" or a "functionalist"? The clearest case is when an individual endorses a universal subordination argument. Such identifications play an important role in the history of biology, where many such arguments have been offered and widely debated. If I am right, however, we should hope that universal subordination arguments will soon be matter *solely* of history – that the future of biology will involve no structuralists or functionalists of this sort.

We also *can* identify structuralists and functionalists based on their endorsement of either restricted subordination arguments or a particular methodological heuristic. In both cases, however, things quickly get complicated. A biologist who accepts a restricted structuralist subordination argument in one domain may accept a restricted functionalist subordination argument in another – that is part of what it means for them to be *restricted*. Likewise, a biologist who promotes methodological structuralism in some cases may promote methodological adaptationism in others. Moreover, one's commitments to any restricted subordination arguments are logically independent of one's methodological commitments. Sure, some individuals may come through the wash with their colors unmixed, identifiable as "structuralist" or "functionalist" in all relevant contexts, but they are likely to be few and far between.

We are none the worse for this. Evolutionary biology neither can nor should want to rid itself of structure/function disputes, but it has no need of structuralists and functionalists.

5 Gentle Polemics

5.1 Stances, Paradigms, and Strategies

I have made the case for analyzing structuralism and functionalism in terms of distinct *explanatory strategies*. I have argued that this approach is able (a) to capture what is shared between diverse structuralist and functionalist figures across an extended historical span, without rendering the positions so abstract as to be explanatorily inert, (b) to explain why structure/function disputes are a recurrent, ineradicable feature of evolutionary biology, while still being empirically tractable, and (c) to explain why the integration of structuralist and functionalist considerations is not only possible, but of comparable importance to more eye-grabbing structure/function disputes. My analysis thus satisfies the five desiderata I identified at the outset (Section 1.3).

In discussing those desiderata, I suggested that my analysis would prove complementary to existing stance (Boucher 2015) and paradigm (Winther 2015) analyses. It is, but the relationship is not so simple as pure complementarity. Any analysis of structuralism and functionalism should be both *explanatory* (why does the history and present of biology look as it does?) and *action-guiding* (how ought we to approach such issues?). In their explanatory roles, I take all three accounts to be compatible and complementary. However, Boucher and I diverge significantly regarding the value of stances: Boucher thinks they drive progress in biology; I think they impede it.

Stances, paradigms, and strategies all have an important role to play in understanding the history and present of biology. Consider stances first. As background value commitments related only nonlogically to empirical evidence, stances help explain why structure/function persist: neither stance can ever be shown to be *wrong*. Further, that a scientist regards certain phenomena as being of especial interest can help explain how they come to hold particular beliefs. Finally, that scientists commit to differing assessments of which phenomena matter can help explain why structure/function disputes often involve incommensurability (e.g. Amundson 2005, chap. 11).

Alongside these virtues, the stance analysis has limits. First, it highlights the nonempirical aspects structure/function disputes at the expense of their empirical aspects. Second, because structuralism and functionalism are incompatible stances, the stance analysis says more about structure/function disputes than about research that integrates structuralist and functionalist considerations. Third, while many biologists have indeed held either a structuralist or functionalist stance, not every contributor to such disputes need do so. One can advance structuralist (functionalist) theorizing without adopting the structuralist (functionalist) stance.

Winther's (2015) paradigm analysis helps overcome these limits. Because paradigms include not merely value commitments, but also technical methods and theoretical views, Winther's analysis can capture the empirical aspects of such research. Moreover, paradigms can relate in many different ways, including both conflict and integration. The major limitation of paradigms is their ephemerality: paradigms are, by design, temporary consolidations of diverse commitments. What do two paradigms, living at different times, have to have in common to both count as structuralist (functionalist)? About this, Winther has little to say.

My account of explanatory strategies helps to tie both of the foregoing accounts together. What do distinct structuralist (functionalist) paradigms have in common? Among their elements, they include structuralist (functionalist) explanatory strategies. How can structuralist and functionalist paradigms be integrated? By developing complex explanatory chains that rely on both structuralist and functionalist strategies. How does a scientist's adopting a particular stance shape their research practices? By directing them to focus on phenomena for which a particular explanatory strategy is most apt. Why can scientists advance structuralist (functionalist) theorizing without adopting a structuralist (functionalist) stance? Because using a particular explanatory strategy does not require endorsing any particular background values. In these ways, consideration of explanatory strategies deepens both the stance and paradigm accounts.

So much for the rose, now for the thorn. I am not satisfied merely to explain why the history of biology looks as it does: that explanation should help us do *better* than our predecessors. It should allow us to more clearly identify the disagreements that animate structure/function disputes and so help us resolve them – or, failing that, at least pursue those disagreements more directly and fruitfully. It should allow us to more readily recognize possible points of nonantagonistic interactions between structuralism and functionalism. And this brings me to my central disagreement with Boucher. This disagreement concerns the question: are stances *good*? Granting that stances are real and that they explain important features of the history of biology, *ought* biologists and philosophers of biology adopt them? Boucher says: yes. I say: no.

This "no" requires clarification. I do not believe that biologists should avoid letting any background value commitments shape their work. I do not believe that they *can*. Every biologist has some value-laden perspective that informs their research. All biologists will "focus on, emphasise, and find meaning and significance in certain features, and de-emphasise others" (Boucher 2015, 391). In this way, a stance can "draw attention to and highlight certain features of the empirical world that are accessible, but harder to see, from other viewpoints" (Boucher 2019, 1373). Thus far, I have no issue.

My worry centers around one particular aspect of the structuralist and functionalist stances, as Boucher (2015) describes them: they involve judgments of which phenomena are *important*. Such judgments are distinct from merely personal fascination with some phenomenon. A biologist fascinated by unity of type and a biologist fascinated by adaptation are different, but they are not (yet) in conflict. A judgment of importance, by contrast, reaches beyond the personal to make a claim about what really matters to evolutionary theorizing:

what must be explained, what is most central to the theory, what lines of research will be most fruitful, and so on (Boucher 2019, 1367–69). Unlike a mere personal preference, a judgment of importance seeks the assent of others: they, too, should share it. Judgments of importance are thus *judgments of taste*, in Kant's (1987 [1790], sec. 32) sense.

My "no" can now be expanded: biologists and philosophers of biology should not endorse either structuralist or functionalist judgments of importance. I have two reasons for opposing such judgments. First, I do not think that such judgments are *necessary*. I do not believe that they are "indispensable in science in guiding scientific work" (as one of my reviewers put it). Second, I do not think that such judgments are *beneficial*. In fact, I think they actively impede progress. I take these reasons up in turn.

Judgments of importance are not required to obtain the benefits of stances. In making certain phenomena especially salient, stances can encourage lines of research that might otherwise be overlooked, and so make certain discoveries more accessible. This is indeed a significant benefit, but it requires only that biologists direct their ingenuity toward those phenomena. What is essential is that those phenomena receive intellectual attention; it is a matter of indifference whether the underlying motivation stems from a judgment of importance or from a merely personal fascination.

Not only can the benefits of stances be obtained without making judgments of importance, such judgments are harmful in three ways. First, there is reason to think that such judgments are by and large just *wrong*. The history of biology makes clear that both structuralists and functionalists have significantly contributed to the advance of biological theorizing. Why think that things are any different *now*? The very structure of evolutionary theory, which provides reason to think that both explanatory strategies are comparably important (Section 4.2), suggests otherwise. Were one or the other judgment of importance to actually succeed in attaining the universal assent it seeks, the cost would be severe. Better, then, if researchers regard their interest in particular phenomena as mere personal preference and act accordingly.

Second, judgments of importance render *even the empirical portions* of structure/function disputes less tractable than they would otherwise be. Questions about which explanatory strategy should be used to explain a particular phenomenon can, with sufficient care, be settled empirically. This requires carefully specifying one's *explananda*, for structuralist and functionalist explanatory strategies are often directed at different aspects of a single phenomenon (Section 2.6). If these differences go unrecognized, incommensurability and miscommunication result. My concern is that judgments of importance can make such incommensurability harder to recognize and resolve, because they will encourage one to regard what is merely *one* way of characterizing a phenomenon as *the* way. Precisely this occurred during early debates evolutionary "constraints"; the result was a great deal of needlessly fruitless debate (Amundson 1994).

Third, judgments of importance promote antagonistic over nonantagonistic interactions. They provide reason to dismiss possibilities of integration. If only certain phenomena and certain types of explanation *really* matter, why investigate how "interesting" phenomena interact with "boring" ones? In fact, however, attending to such interactions significantly improves evolutionary explanations (Section 3). The biological world does not sort neatly into "the domain where structuralist explanations work" and "the domain where functionalist explanations work." Evolutionary theorizing that treats one or the other as unimportant is going to be impoverished *even in its home domain*: functionalists who neglect structuralist explanations will be worse functionalists (Section 3.2), and structuralists who neglect functionalist explanations will be worse structuralists (Section 3.3).

I therefore hope, if perhaps in vain, for a future biology free of structuralist and functionalist stances. Evolutionary biology needs structuralist and functionalist explanations, and even structure/function disputes, but – if I might repeat myself – it has no need of structuralists and functionalists.

5.2 Is a New and General Theory of Evolution Emerging?

I have made the case for approaching ongoing disputes in evolutionary biology with an emphasis on explanatory strategies, hoping that doing so will allow us to leave behind intractable conflicts between stances. By their nature, however, stances usually conflict indirectly, through the particular theories that serve as their proxies. I would therefore like to consider how focusing on explanatory strategies can illuminate conflict between these proxies. My basic suggestion is that it will let us forgo that great shibboleth "evolutionary theory" – though not, to be sure, evolutionary theorizing.

The rise of evo-devo followed a fairly common trajectory. Early work presented itself as a radical challenge to the mainstream (e.g., Alberch 1989) and received pushback accordingly (e.g., Reeve and Sherman 1993). Possibly this served an important rhetorical function, helping the new science establish its importance. As evo-devo found surer footing, however, many of these allegedly deep conflicts melted away – through empirical discoveries about the operation of gene regulatory networks (Peter and Davidson 2015), through work integrating theories and methods in evo-devo with mainstream theories and methods (Arthur 2001; Nunes et al. 2013), through conceptual work disentangling various points of incommensurability (Amundson 1994; Lewens 2009; Novick 2019), and through institutional entrenchment that allowed evo-devo to be less anxious about its status.

This trajectory is not the result of any special features of evo-devo. Rather, it is highly *predictable* (Mitchell and Dietrich 2006). Disputes in biology frequently begin as high-level conflicts between competing "theories" or "paradigms" or "frameworks" or what-have-you, only to eventually shift to a more conciliatory relationship founded on (a) a clearer understanding of distinct explanatory domains (still allowing for local conflicts) and (b) the success of attempts to integrate across these domains. We see it in the shift from rancorous debates between neutralists and selectionists to the general acceptance of neutral models of evolution, both as powerful explanatory tools in their own right and as methodological resources for selectionists (Dietrich 2006). We see it, too, in the rise of systems biology: what began as a self-conceived competitor to molecular biology has now largely made peace with its former enemy, allowing both to stably coexist in predominantly sequestered domains (Gross, Kranke, and Meunier 2019).

Why do so many conflicts in biology follow similar trajectories? The reasons lie, at least partially, in the nature of the material that biologists study. Organisms are both *complex* and *evolved*. Because they are complex, it is unlikely – no matter what aspect of organisms one studies – that a single approach will answer all of the why-questions worth asking (Mitchell 2009; Mitchell and Gronenborn 2017). Because they are evolved, biological theorizing must cover a remarkably heterogeneous domain, shaped by contingent historical processes (Beatty 1995). Biology is just not the right domain for constructing grand unified theories. This is especially salient in evolutionary contexts: evolutionary processes have a material basis (the organisms, populations, and lineages on which they act), and this material basis is itself evolved. The evolutionary processes that dominate in one lineage, or at one time in a given lineage's history, or with regard to one aspect of a single lineage, need not dominate in other lineages, at other times, or with regard to other aspects. The diversity of life includes the diversity of processes by which life evolves.

In light of this, it is a fool's errand to analyze theoretical conflicts in biology as occurring between entrenched general theories and the upstart, equally general theories that seek to replace them. A growing body of work in the philosophy of biology, focusing on a range of debates, has coalesced around this view (Love 2013, forthcoming, sec. 1.3; Booth, Mariscal, and Doolittle 2016; Buskell 2019, 2020; Lewens 2019; Novick and Doolittle 2019; Reis and Araújo 2020; Svensson 2021). To Gould's (1980) classic question – "is a new and general theory of evolution emerging" – the answer is always *no*, not because existing theorizing is satisfactory, but because there is neither a new and general theory emerging nor an old and general theory for it to replace.

Within biology, however, rhetoric of conflict between competing general theories of evolution is common. Often, this takes the form of setting up "the modern synthetic theory of evolution" as an ossified bogeyman, characterized by a distinct set of assumptions and claims, that needs to be replaced (e.g., Gould 1980; Pigliucci 2007; Laland et al. 2015; see Svensson 2021 for further examples). Such an approach invariably falsifies the history of the modern synthesis (Svensson 2021), which might be forgivable, but in doing so it obscures what is really at stake in the contemporary debates, which – with apologies to my historian friends, including her who lives in my head – I find more difficult to pardon.

Much better, then, to approach theoretical debates in biology with a focus on explanatory strategies. The aim should be the clear articulation of distinct explanatory strategies, with an eye toward understanding (a) their proper domains of application, (b) their possibilities for integration (which are much easier to see with this low-level focus), and, yes, (c) the ways in which they come into genuine conflict. Such conflicts, however, are not conflicts between distinct explanatory strategies as such, but rather are conflicts concerning how those strategies apply in particular instances.

It is not hard to surmise that, in writing this section, I have in mind recent calls for an "extended evolutionary synthesis" (Laland et al. 2015; Müller 2017). On the approach I am recommending, the extended synthesis should be viewed as a loose-knit collection of explanatory strategies, each of which stands or falls on its own, but which may be locally integrated both with each other and with older explanatory strategies. Rhetoric surrounding the extended synthesis vacillates between conflict-first and integration-first modes; I regard the former as tendentious and obfuscatory and the latter as much more promising. My point, however, is not about the extended synthesis as such. It is the current rallying point for critics of the evolutionary mainstream, but it, too, will eventually make its peace with that mainstream, one way or another. It is the latest, but surely not the last, link in a long chain of debates, and I am happy to pre-emptively raise the same criticism of, and offer the same advice to, whatever comes next.

5.3 Fleabane and Wormwood

This Element, which winds now to its close, is many things. It is a philosophical analysis of structuralism and functionalism. It is an account of the internal conceptual structure of evolutionary-developmental biology. It is an attempt to resolve, or at least indicate how one *might* resolve, ongoing disputes within evolutionary theory. But it is also, below the surface, a methodological plea for biologists and philosophers of biology, a plea to take our debates smaller and more local, away from stances and general theories and toward explanatory strategies. And this puts me in an awkward position, for this plea itself emerges from a stance, from my own sense, as a philosopher of biology, of which questions most matter and which modes of inquiry will most insightfully answer them. This stance I call *Zhuangist*, after the Warring States philosopher Zhuang Zhou (Ziporyn 2020).

Justifications of stances are always pragmatic, never definitive. What justification of the Zhuangist stance I have to offer, I have offered already. Adopting this stance has led me to develop an analysis of structuralism and functionalism that – if my arguments have been successful – clarifies and helps explain biology's long history of structure/function disputes (Section 2), does justice to the internal complexities of evo-devo (Section 3), is consistent with the nature of evolutionary theorizing (Section 4), and points the way forward in areas of ongoing controversy (Section 5). These exhaust my public reasons for adopting it. But a book is a personal affair, so perhaps it is permissible to end with some private history.

This Element's origins lie in a confrontation that occurred in autumn 2012, my first term in graduate school, when I took a seminar in the philosophy of biology taught by James Lennox. Jim had us read Ron Amundson's (2005) fascinating analysis of evo-devo. This proved a shock to the "neo-Darwinian" (close enough) understanding of evolutionary theory I had formed long before, as a child who liked catching snakes and arguing with creationists – an understanding sophisticated but largely unchallenged by my undergraduate studies. The Element set my mind aflame, upsetting my sense of "how evolution works." It set me aflame, but – I did not understand it. How, exactly, did evo-devo's structuralist explanations work? How could they be so successful despite conflicting so deeply with the functionalism of the evolutionary mainstream?

That original confusion is the fount of most of my subsequent intellectual labor. Unpacking Amundson's analysis of evo-devo, I found that my difficulties lay, not with the details, but with his approach: his identification of two separate, divided, internally unified research programs and associated theoretical edifices. The more I poked around, the more I found *both* sources of internal complexity and disunity *and* points of fruitful integration (Novick 2018, 2019). Empirical conflicts remained, but the conceptual conflict, the impassable incommensurability with which Amundson's book ends, came to seem like an artifact of the mode of presentation.

So it has gone for me generally. Whenever I encounter debates about competing general visions of evolutionary theory, I feel lost at sea, unsure not only of *how* such a clash might be resolved, but even of what it would *mean* to resolve it. In trying to ease this discomfort, to relocate the shore, what I find succeeds, time and again, is to isolate local conflicts over how to explain particular phenomena. I know what it is to resolve *these*, whether empirically, by obtaining the right evidence, or conceptually, by disentangling subtly distinct *explananda* that have been conflated.

This, then, is my private reason for adopting the Zhuangist stance: it alone has allowed me to ease the perplexity from which inquiry springs. Intractable conflict strikes me as always and everywhere a *mistake*, as a sign that something remains not yet understood. Only once I have found and cleared a path between two distinct perspectives do I feel intellectually at ease. And once that local footpath is passable, I find that the larger conflicts simply fall away: nothing more seems to turn on them.

Perhaps this notion – that, with sufficient attention to local possibilities of integration, large-scale conflicts will vanish – is soft and doe-eyed. Perhaps I underestimate both the stubbornness and the value of such conflicts. Perhaps large-scale commitments are needed to give small-scale activities sense and motivation, and so perhaps their disputes are fruitful even if irresolvable. Consider Emerson (1983 [1844]):

> We aim above the mark, to hit the mark. Every act hath some falsehood of exaggeration in it. And when now and then comes along some sad, sharp-eyed man, who sees how paltry a game is played, and refuses to play, but blabs the secret; – how then? is the bird flown? O no, the wary Nature sends a new troop of fairer forms, of lordlier youths, with a little more excess of direction to hold them fast to their several aim; makes them a little wrong-headed in that direction in which they are rightest, and on goes the game again with new whirl, for a generation or two more.

Am I playing the sad, sharp-eyed woman? I hope not; I think not. Why, so long as developing and applying explanatory strategies yields answers to our questions, need we burden this activity with commitments that – it seems to me – do little more than generate intractable conflict? Why should the whole-hearted pursuit of our interests require not just partiality – that ineluctable fate of all sentient beings – but *exclusive* partiality? I see no reason.

But I am, at last, depleted even of private reasons. A story, then, taken from the first chapter of the *Zhuangzi*, the finest book I know (my translation):

> 窮髮之北，有冥海者，天池也。有魚焉，其廣數千里，未有知其脩者，其名為鯤。有鳥焉，其名為鵬，背若泰山，翼若垂天之雲，摶扶搖羊角而上者九萬里，絕雲氣，負青天，然後圖南，且適南冥也。斥鴳笑之曰:「彼且奚適也? 我騰躍而上，不過數仞而下，翱翔蓬蒿之間，此亦飛之至也。而彼且奚適也，」

> In the desolate north is a deep ocean – Heaven's Pond. There is a fish there, thousands of *li* wide, its length unknown – its name is Roe. There is a bird there – its name is Roc, with a back like Mount Tai, wings like Heaven's hanging clouds. Spiraling up a goat's horn cyclone and climbing 90,000 *li*, vanishing in clouds, bearing blue Heaven on its back, then charting a course south – soon it will reach the southern deep. The scold-quail laughs at it, saying,
>
> *Where is* that *going?*
> *I bounce and rise,*
> *cover but few fathoms and fall,*
> *take wing and soar*
> *amid fleabane and wormwood –*
> this *is flying's utmost.*
> *So where is* that *going?*

Who could object to the quail's flight amid fleabane and wormwood? Such flight suits the little bird's capacities, enabling many discoveries and delights – delights inaccessible to the unfathomably huge Roc, who can fly only at a height so great that all happenings on the earth below are entirely indiscernible. No, the low-lying flight of the quail is above reproach. But to be a *scold*-quail, to call such cavorting *flying's utmost*, to see nothing of value beyond fleabane and wormwood – sad, isn't it!

References

Alberch, Pere. 1989. “The Logic of Monsters: Evidence for Internal Constraint in Development and Evolution.” *Geobios*, Ontogenèse Et Évolution 22 (January): 21–57. https://doi.org/10.1016/S0016-6995(89)80006-3.

Alberch, Pere, and Emily A. Gale. 1985. “A Developmental Analysis of an Evolutionary Trend: Digital Reduction in Amphibians.” *Evolution* 39 (1): 8–23. https://doi.org/10.1111/j.1558-5646.1985.tb04076.x.

Amundson, Ron. 1994. “Two Concepts of Constraint: Adaptationism and the Challenge from Developmental Biology.” *Philosophy of Science* 61 (4): 556–78. https://doi.org/10.1086/289822.

2005. *The Changing Role of the Embryo in Evolutionary Thought: Roots of Evo-Devo*. Cambridge Studies in Philosophy and Biology. Cambridge: Cambridge University Press. https://doi.org/10.1017/CBO9781139164856.

2021. “Form and Function in Evo-Devo.” In *Evolutionary Developmental Biology*, edited by Laura Nuño de la Rosa and Gerd B. Müller, 457–67. Cham: Springer International. https://doi.org/10.1007/978-3-319-32979-6_91.

Andersson, Jan O. 2005. “Lateral Gene Transfer in Eukaryotes.” *Cellular and Molecular Life Sciences* 62 (11): 1182–97. https://doi.org/10.1007/s00018-005-4539-z.

Appel, Toby A. 1987. *The Cuvier-Geoffrey Debate: French Biology in the Decades before Darwin*. New York: Oxford University Press.

Arias Del Angel, Juan A., Vidyanand Nanjundiah, Mariana Benítez, and Stuart A. Newman. 2020. “Interplay of Mesoscale Physics and Agent-Like Behaviors in the Parallel Evolution of Aggregative Multicellularity.” *EvoDevo* 11 (1): 21. https://doi.org/10.1186/s13227-020-00165-8.

Arthur, Wallace. 2000. *The Origin of Animal Body Plans: A Study in Evolutionary Developmental Biology*. Cambridge, UK: Cambridge University Press.

2001. “Developmental Drive: An Important Determinant of the Direction of Phenotypic Evolution.” *Evolution & Development* 3 (4): 271–78. https://doi.org/10.1046/j.1525-142x.2001.003004271.x.

2002a. “The Emerging Conceptual Framework of Evolutionary Developmental Biology.” *Nature* 415 (6873): 757–64. https://doi.org/10.1038/415757a.

2002b. “The Interaction between Developmental Bias and Natural Selection: From Centipede Segments to a General Hypothesis.” *Heredity* 89 (4): 239–46. https://doi.org/10.1038/sj.hdy.6800139.

2004. "The Effect of Development on the Direction of Evolution: Toward a Twenty-First Century Consensus." *Evolution & Development* 6 (4): 282–88. https://doi.org/10.1111/j.1525-142X.2004.04033.x.

2015. "Internal Factors in Evolution: The Morphogenetic Tree, Developmental Bias, and Some Thoughts on the Conceptual Structure of Evo-Devo." In *Conceptual Change in Biology: Scientific and Philosophical Perspectives on Evolution and Development*, edited by Alan C. Love, 343–63. Boston Studies in the Philosophy and History of Science. Dordrecht: Springer Netherlands. https://doi.org/10.1007/978-94-017-9412-1_16.

2021. *Understanding Evo-Devo*. Cambridge: Cambridge University Press.

Asma, Stephen T. 1996. *Following Form and Function: A Philosophical Archaeology of Life Science*. Evanston: Northwestern University Press.

Austin, Christopher J. 2017. "Evo-Devo: A Science of Dispositions." *European Journal for Philosophy of Science* 7 (2): 373–89. https://doi.org/10.1007/s13194-016-0166-9.

Barrett, Paul H., Peter J. Gautrey, Sandra Herbert, David Kohn, and Sidney Smith, eds. 2009. *Charles Darwin's Notebooks, 1836–1844: Geology, Transmutation of Species, Metaphysical Enquiries*. Cambridge: Cambridge University Press.

Beatty, John. 1995. "The Evolutionary Contingency Thesis." In *Concepts, Theories, and Rationality in the Biological Sciences*, edited by Gereon Wolters and James G. Lennox, 45–81. Pittsburgh: University of Pittsburgh Press.

2016. "The Creativity of Natural Selection? Part I: Darwin, Darwinism, and the Mutationists." *Journal of the History of Biology* 49 (4): 659–84. https://doi.org/10.1007/s10739-016-9456-5.

2019. "The Creativity of Natural Selection? Part II: The Synthesis and Since." *Journal of the History of Biology* 52: 705–31. https://doi.org/10.1007/s10739-019-09583-4.

Benítez, Mariana, Valeria Hernández-Hernández, Stuart A. Newman, and Karl J. Niklas. 2018. "Dynamical Patterning Modules, Biogeneric Materials, and the Evolution of Multicellular Plants." *Frontiers in Plant Science* 9: 871, 1–16. www.frontiersin.org/articles/10.3389/fpls.2018.00871.

Booth, Austin, Carlos Mariscal, and W. Ford Doolittle. 2016. "The Modern Synthesis in the Light of Microbial Genomics." *Annual Review of Microbiology* 70 (July): 279–97. https://doi.org/10.1146/annurev-micro-102215-095456.

Borkin, Leo J., and Mikhail M. Pikulik. 1986. "The Occurrence of Polymely and Polydactyly in Natural Populations of Anurans of the USSR." *Amphibia-Reptilia* 7 (3): 205–16. https://doi.org/10.1163/156853886X00019.

Boucher, Sandy C. 2014. "What is a Philosophical Stance? Paradigms, Policies and Perspectives." *Synthese* 191 (10): 2315–32. https://doi.org/10.1007/s11229-014-0400-y.

2015. "Functionalism and Structuralism as Philosophical Stances: Van Fraassen Meets the Philosophy of Biology." *Biology & Philosophy* 30 (3): 383–403. https://doi.org/10.1007/s10539-014-9453-z.

2019. "An Empiricist Conception of the Relation between Metaphysics and Science." *Philosophia* 47 (5): 1355–78. https://doi.org/10.1007/s11406-018-0040-4.

Breuker, Casper J., Vincent Debat, and Christian Peter Klingenberg. 2006. "Functional Evo-Devo." *Trends in Ecology & Evolution* 21 (9): 488–92. https://doi.org/10.1016/j.tree.2006.06.003.

Brigandt, Ingo. 2021. "Typology and Natural Kinds in Evo-Devo." In *Evolutionary Developmental Biology: A Reference Guide*, edited by Laura Nuño de la Rosa and Gerd B. Müller, 483–93. Cham: Springer International. https://doi.org/10.1007/978-3-319-32979-6_100.

Brigandt, Ingo, and Alan C. Love. 2010. "Evolutionary Novelty and the Evo-Devo Synthesis: Field Notes." *Evolutionary Biology* 37 (2): 93–99. https://doi.org/10.1007/s11692-010-9083-6.

2012. "Conceptualizing Evolutionary Novelty: Moving Beyond Definitional Debates." *Journal of Experimental Zoology Part B: Molecular and Developmental Evolution* 318 (6): 417–27. https://doi.org/10.1002/jez.b.22461.

Brown, Rachael L. In Press. "Structuralism and Adaptationism: Friends? Or Foes?" *Seminars in Cell & Developmental Biology*. https://doi.org/10.1016/j.semcdb.2022.02.022.

Brunet, Tyler D. P. 2022. "Higher Level Constructive Neutral Evolution." *Biology & Philosophy* 37 (4): 23, 1–22. https://doi.org/10.1007/s10539-022-09858-x.

Brunet, T. D. P., and W. Ford Doolittle. 2018. "The Generality of Constructive Neutral Evolution." *Biology & Philosophy* 33 (1): 2, 1–25. https://doi.org/10.1007/s10539-018-9614-6.

Buskell, Andrew. 2019. "Reciprocal Causation and the Extended Evolutionary Synthesis." *Biological Theory* 14: 267–79. https://doi.org/10.1007/s13752-019-00325-7.

2020. "Synthesising Arguments and the Extended Evolutionary Synthesis." *Studies in History and Philosophy of Science Part C: Studies in History and Philosophy of Biological and Biomedical Sciences* 80: 101244. https://doi.org/10.1016/j.shpsc.2019.101244.

Camardi, Giovanni. 2001. "Richard Owen, Morphology and Evolution." *Journal of the History of Biology* 34 (3): 481–515. https://doi.org/10.1023/A:1012946930695.

Carroll, Sean B. 2005. "Evolution at Two Levels: On Genes and Form." *PLOS Biology* 3 (7): e245. https://doi.org/10.1371/journal.pbio.0030245.

2008. "Evo-Devo and an Expanding Evolutionary Synthesis: A Genetic Theory of Morphological Evolution." *Cell* 134 (1): 25–36. https://doi.org/10.1016/j.cell.2008.06.030.

Chung, Carl. 2003. "On the Origin of the Typological/Population Distinction in Ernst Mayr's Changing Views of Species, 1942–1959." *Studies in History and Philosophy of Science Part C: Studies in History and Philosophy of Biological and Biomedical Sciences* 34 (2): 277–96. https://doi.org/10.1016/S1369-8486(03)00026-8.

Coleman, William. 1964. *Georges Cuvier, Zoologist: A Study in the History of Evolution Theory*. Cambridge, MA: Harvard University Press.

1977 (1971). *Biology in the Nineteenth Century: Problems of Form, Function, and Transformation*. Cambridge: Cambridge University Press.

Coyne, Jerry A. 2006. "Comment on 'Gene Regulatory Networks and the Evolution of Animal Body Plans.'" *Science* 313 (5788): 761. https://doi.org/10.1126/science.1126454.

Craig, Lindsay R. 2009. "Defending Evo-Devo: A Response to Hoekstra and Coyne." *Philosophy of Science* 76 (3): 335–44. https://doi.org/10.1086/649808.

Darwin, Charles. 1964 (1859). *On the Origin of Species: A Facsimile of the First Edition*. Facsimile ed. Cambridge, MA: Harvard University Press.

Davidson, Eric H., and Douglas H. Erwin. 2006. "Gene Regulatory Networks and the Evolution of Animal Body Plans." *Science* 311 (5762): 796–800. https://doi.org/10.1126/science.1113832.

Dawkins, Richard. 1986. *The Blind Watchmaker: Why the Evidence of Evolution Reveals a Universe without Design*. New York: Norton.

de Robertis, Edward M. 2008. "The Molecular Ancestry of Segmentation Mechanisms." *Proceedings of the National Academy of Sciences* 105 (43): 16411–12. https://doi.org/10.1073/pnas.0808774105.

Dietrich, Michael R. 2006. "Three Perspectives on Neutrality and Drift in Molecular Evolution." *Philosophy of Science* 73 (5): 666–77. https://doi.org/10.1086/518521.

DiFrisco, James, Alan C. Love, and Günter P. Wagner. 2020. "Character Identity Mechanisms: A Conceptual Model for Comparative-Mechanistic Biology." *Biology & Philosophy* 35 (4): 44, 1–32. https://doi.org/10.1007/s10539-020-09762-2.

DiFrisco, James, Günter P. Wagner, and Alan C. Love. In Press. "Reframing Research on Evolutionary Novelty and Co-Option: Character Identity Mechanisms versus Deep Homology." *Seminars in Cell & Developmental Biology*. https://doi.org/10.1016/j.semcdb.2022.03.030.

Dobzhansky, Theodosius. 1982 (1937). *Genetics and the Origin of Species*. New York: Columbia University Press.

Doolittle, W. Ford. 2012. "A Ratchet for Protein Complexity." *Nature* 481 (7381): 270–71. https://doi.org/10.1038/nature10816.

Duboule, Denis. 2007. "The Rise and Fall of Hox Gene Clusters." *Development* 134 (14): 2549–60. https://doi.org/10.1242/dev.001065.

Emerson, Ralph Waldo. 1983 (1844). "Nature (1844)." In *Essays and Lectures*, edited by Joel Porte, 539–55. New York: Library of America.

Endler, John A. 1986. *Natural Selection in the Wild*. Princeton: Princeton University Press.

Erwin, Douglas H. 2020. "The Origin of Animal Body Plans: A View from Fossil Evidence and the Regulatory Genome." *Development* 147 (4): dev182899, 1–14. https://doi.org/10.1242/dev.182899.

Erwin, Douglas H., and James Valentine. 2013. *The Cambrian Explosion: The Construction of Animal Biodiversity*. 1st ed. Greenwood Village: Roberts.

Fábregas-Tejeda, Alejandro, and Francisco Vergara-Silva. 2018. "Hierarchy Theory of Evolution and the Extended Evolutionary Synthesis: Some Epistemic Bridges, Some Conceptual Rifts." *Evolutionary Biology* 45 (2): 127–39. https://doi.org/10.1007/s11692-017-9438-3.

Fusco, Giuseppe. 2015. "For a New Dialogue between Theoretical and Empirical Studies in Evo-Devo." *Frontiers in Ecology and Evolution* 3: 97, 1–6. https://doi.org/10.3389/fevo.2015.00097.

Garson, Justin. 2017. "A Generalized Selected Effects Theory of Function." *Philosophy of Science* 84 (3): 523–43. https://doi.org/10.1086/692146.

Gilbert, Scott F. 2016. "Developmental Plasticity and Developmental Symbiosis: The Return of Eco-Devo." In *Current Topics in Developmental Biology*, edited by Paul M. Wassarman, 116: 415–33. Essays on Developmental Biology, Part A. Cambridge, MA: Academic Press. https://doi.org/10.1016/bs.ctdb.2015.12.006.

Godfrey-Smith, Peter. 2001. "Three Kinds of Adaptationism." In *Adaptationism and Optimality*, edited by Steven Hecht Orzack and Elliott Sober, 335–57. Cambridge: Cambridge University Press.

2009. *Darwinian Populations and Natural Selection*. Oxford: Oxford University Press.

Goldschmidt, Richard. 1982 (1940). *The Material Basis of Evolution*. New Haven: Yale University Press.

Gould, Stephen J. 1980. "Is a New and General Theory of Evolution Emerging?" *Paleobiology* 6 (1): 119–30. https://doi.org/10.1017/S0094837300012549.

Gould, Stephen J., and Richard C. Lewontin. 1979. "The Spandrels of San Marco and the Panglossian Paradigm: A Critique of the Adaptationist Programme." *Proceedings of the Royal Society B* 205 (1161): 581–98. https://doi.org/10.1098/rspb.1979.0086.

Gray, Michael W., Julius Lukeš, John M. Archibald, Patrick J. Keeling, and W. Ford Doolittle. 2010. "Irremediable Complexity?" *Science* 330 (6006): 920–21. https://doi.org/10.1126/science.1198594.

Greer, Joy M., John Puetz, Kirk R. Thomas, and Mario R. Capecchi. 2000. "Maintenance of Functional Equivalence during Paralogous Hox Gene Evolution." *Nature* 403 (6770): 661–65. https://doi.org/10.1038/35001077.

Gross, Fridolin, Nina Kranke, and Robert Meunier. 2019. "Pluralization through Epistemic Competition: Scientific Change in Times of Data-Intensive Biology." *History and Philosophy of the Life Sciences* 41 (1): 1, 1–29. https://doi.org/10.1007/s40656-018-0239-5.

Guzmán-Herrera, Alejandra, Juan A. Arias Del Angel, Natsuko Rivera-Yoshida, Mariana Benítez, and Alessio Franci. 2021. "Dynamical Patterning Modules and Network Motifs as Joint Determinants of Development: Lessons from an Aggregative Bacterium." *Journal of Experimental Zoology Part B: Molecular and Developmental Evolution* 336 (3): 300–14. https://doi.org/10.1002/jez.b.22946.

Hall, Brian K. 2000. "Guest Editorial: Evo-Devo or Devo-Evo – Does It Matter?" *Evolution & Development* 2 (4): 177–78. https://doi.org/10.1046/j.1525-142x.2000.00003e.x.

Hernández-Hernández, Valeria, Karl J. Niklas, Stuart A. Newman, and Mariana Benítez. 2012. "Dynamical Patterning Modules in Plant Development and Evolution." *International Journal of Developmental Biology* 56 (9): 661–74. https://doi.org/10.1387/ijdb.120027mb.

Ho, Mae-Wan, and Peter T. Saunders. 1993. "Rational Taxonomy and the Natural System." *Acta Biotheoretica* 41 (4): 289–304. https://doi.org/10.1007/BF00709367.

Hoekstra, Hopi E., and Jerry A. Coyne. 2007. "The Locus of Evolution: Evo Devo and the Genetics of Adaptation." *Evolution* 61 (5): 995–1016. https://doi.org/10.1111/j.1558-5646.2007.00105.x.

Hughes, Anthony J., and David M. Lambert. 1984. "Functionalism, Structuralism, and 'Ways of Seeing.'" *Journal of Theoretical Biology* 111 (4): 787–800. https://doi.org/10.1016/S0022-5193(84)80267-2.

Hull, David L. 1965a. "The Effect of Essentialism on Taxonomy: Two Thousand Years of Stasis (I)." *The British Journal for the Philosophy of Science* 15 (60): 314–26. https://doi.org/10.1093/bjps/XV.60.314.

1965b. "The Effect of Essentialism on Taxonomy: Two Thousand Years of Stasis (II)." *The British Journal for the Philosophy of Science* 16 (61): 1–18.

Irwin, Terence, and Gail Fine, trans. 1996. *Aristotle: Introductory Readings*. Indianapolis: Hackett.

Jenner, Ronald A. 2006. "Unburdening Evo-Devo: Ancestral Attractions, Model Organisms, and Basal Baloney." *Development Genes and Evolution* 216 (7): 385–94. https://doi.org/10.1007/s00427-006-0084-5.

Joyce, James. 2012 (1939). *Finnegans Wake*. Revised ed. Oxford: Oxford University Press.

Kant, Immanuel. 1987 (1790). *Critique of Judgment*. Translated by Werner S. Pluhar. Indianapolis: Hackett.

Kauffman, Stuart A. 1993. *The Origins of Order: Self-Organization and Selection in Evolution*. Oxford: Oxford University Press.

Kimura, Motoo. 1983. *The Neutral Theory of Molecular Evolution*. Cambridge: Cambridge University Press. https://doi.org/10.1017/CBO9780511623486.

Laland, Kevin N., Tobias Uller, Marcus W. Feldman et al. 2015. "The Extended Evolutionary Synthesis: Its Structure, Assumptions and Predictions." *Proceedings of the Royal Society B: Biological Sciences* 282 (1813): 20151019. https://doi.org/10.1098/rspb.2015.1019.

Lamarck, Jean Baptiste Pierre Antoine de Monet de. 2011 (1809). *Zoological Philosophy: An Exposition with Regard to the Natural History of Animals*. Translated by Hugh Samuel Roger Elliott. Cambridge: Cambridge University Press. https://doi.org/10.1017/CBO9781139105323.

Laubichler, Manfred D. 2009. "Form and Function in Evo Devo: Historical and Conceptual Reflections." In *Form and Function in Developmental Evolution*, edited by Jane Maienschein and Manfred D. Laubichler, 10–46. Cambridge Studies in Philosophy and Biology. Cambridge: Cambridge University Press. https://doi.org/10.1017/CBO9780511576188.002.

Lemons, Derek, and William McGinnis. 2006. "Genomic Evolution of Hox Gene Clusters." *Science* 313 (5795): 1918–22. https://doi.org/10.1126/science.1132040.

Lennox, James G. 2021. "Form as Cause and the Formal Cause: Aristotle's Answer." In *Neo-Aristotelian Perspectives on Formal Causation*, edited by Ludger Jansen and Petter Sandstad, 225–37. New York: Routledge.

Lewens, Tim. 2008. "Seven Types of Adaptationism." *Biology & Philosophy* 24 (2): 161, 1–22. https://doi.org/10.1007/s10539-008-9145-7.

2009. "Evo-Devo and 'Typological Thinking': An Exculpation." *Journal of Experimental Zoology Part B: Molecular and Developmental Evolution* 312B (8): 789–96. https://doi.org/10.1002/jez.b.21292.

2019. "The Extended Evolutionary Synthesis: What is the Debate about, and What Might Success for the Extenders Look Like?" *Biological Journal of the Linnean Society* 127 (4): 707–21. https://doi.org/10.1093/biolinnean/blz064.

Lewontin, Richard C. 1983. "Gene, Organism, and Environment." In *Evolution from Molecules to Men*, edited by Derek S. Bendall, 273–85. Cambridge: Cambridge University Press.

Linde-Medina, Marta. 2010. "Two 'EvoDevos.'" *Biological Theory* 5 (1): 7–11. https://doi.org/10.1162/BIOT_a_00014.

Lloyd, Elisabeth A. 2015. "Adaptationism and the Logic of Research Questions: How to Think Clearly About Evolutionary Causes." *Biological Theory* 10 (4): 343–62. https://doi.org/10.1007/s13752-015-0214-2.

Longabaugh, William J. R., Eric H. Davidson, and Hamid Bolouri. 2005. "Computational Representation of Developmental Genetic Regulatory Networks." *Developmental Biology* 283 (1): 1–16. https://doi.org/10.1016/j.ydbio.2005.04.023.

2009. "Visualization, Documentation, Analysis, and Communication of Large-Scale Gene Regulatory Networks." *Biochimica et Biophysica Acta (BBA) – Gene Regulatory Mechanisms*, System Biology – Genetic Networks 1789 (4): 363–74. https://doi.org/10.1016/j.bbagrm.2008.07.014.

Love, Alan C. 2013. "Theory is as Theory Does: Scientific Practice and Theory Structure in Biology." *Biological Theory* 7 (4): 325–37. https://doi.org/10.1007/s13752-012-0046-2.

forthcoming. *Evolution and Development: Conceptual Issues*. Elements in the Philosophy of Biology. Cambridge: Cambridge University Press.

Lu, Qiaoying, and Pierrick Bourrat. 2018. "The Evolutionary Gene and the Extended Evolutionary Synthesis." *The British Journal for the Philosophy of Science* 69 (3): 775–800. https://doi.org/10.1093/bjps/axw035.

Lukeš, Julius, John M. Archibald, Patrick J. Keeling, W. Ford Doolittle, and Michael W. Gray. 2011. "How a Neutral Evolutionary Ratchet Can Build Cellular Complexity." *IUBMB Life* 63 (7): 528–37. https://doi.org/10.1002/iub.489.

Mayr, Ernst. 1959. "Darwin and the Evolutionary Theory in Biology." In *Evolution and Anthropology: A Centennial Appraisal*, edited by Betty J. Meggers, 1–10. Washington, DC: Anthropological Society of Washington.

1960. "The Emergence of Evolutionary Novelties." In *Evolution after Darwin, Vol. 1, The Evolution of Life: Its Origin, History, and Future*, edited by Sol Tax, 349–80. Chicago: University of Chicago Press.

1969. "Footnotes on the Philosophy of Biology." *Philosophy of Science* 36 (2): 197–202. https://doi.org/10.1086/288246.

1982 (1942). *Systematics and the Origin of Species: The Columbia Classics in Evolution Series*. New York: Columbia University Press.

Mayr, Ernst, and William B. Provine. 1981. "The Evolutionary Synthesis." *Bulletin of the American Academy of Arts and Sciences* 34 (8): 17–32. https://doi.org/10.2307/3823367.

McConwell, Alison K. 2019. "Contingency's Causality and Structural Diversity." *Biology & Philosophy* 34 (2): 26, 1–26. https://doi.org/10.1007/s10539-019-9679-x.

McShea, Daniel W. 1991. "Complexity and Evolution: What Everybody Knows." *Biology and Philosophy* 6 (3): 303–24. https://doi.org/10.1007/BF00132234.

Minchin, Edward A. 1898. "Materials for a Monograph of the Ascons: I. On the Origin and Growth of the Triradiate and Quadriradiate Spicules in the Family Clathrinidæ." *Journal of Cell Science* s2-40 (160): 469–587.

Mitchell, Sandra D. 2009. *Unsimple Truths*. Chicago: University of Chicago Press. www.degruyter.com/document/doi/10.7208/9780226532653/html.

Mitchell, Sandra D., and Angela M. Gronenborn. 2017. "After Fifty Years, Why Are Protein X-Ray Crystallographers Still in Business?" *The British Journal for the Philosophy of Science* 68 (3): 703–23. https://doi.org/10.1093/bjps/axv051.

Mitchell, Sandra D., and Michael R. Dietrich. 2006. "Integration without Unification: An Argument for Pluralism in the Biological Sciences." *The American Naturalist* 168 (S6): S73–79. https://doi.org/10.1086/509050.

Mossio, Matteo, Cristian Saborido, and Alvaro Moreno. 2009. "An Organizational Account of Biological Functions." *The British Journal for the Philosophy of Science* 60 (4): 813–41. https://doi.org/10.1093/bjps/axp036.

Müller, Gerd B. 2017. "Why an Extended Evolutionary Synthesis is Necessary." *Interface Focus* 7 (5): 20170015. https://doi.org/10.1098/rsfs.2017.0015.

Muñoz-Gómez, Sergio A., Gaurav Bilolikar, Jeremy G. Wideman, and Kerry Geiler-Samerotte. 2021. "Constructive Neutral Evolution 20 Years Later." *Journal of Molecular Evolution* 89 (3): 172–82. https://doi.org/10.1007/s00239-021-09996-y.

Newman, Stuart A. 2016. "'Biogeneric' Developmental Processes: Drivers of Major Transitions in Animal Evolution." *Philosophical Transactions of the Royal Society of London. Series B, Biological Sciences* 371 (1701): 20150443. https://doi.org/10.1098/rstb.2015.0443.

2021. "The Origins and Evolution of Animal Identity." In *Biological Identity: Perspectives from Metaphysics and the Philosophy of Biology*, edited by Anne Sophie Meincke and John Dupré, 128–48. London: Routledge.

2022. "Form, Function, Agency: Sources of Natural Purpose in Animal Evolution." *EcoEvoRxiv*. https://doi.org/10.32942/osf.io/szv35.

Newman, Stuart A., and Ramray Bhat. 2008. "Dynamical Patterning Modules: Physico-Genetic Determinants of Morphological Development and Evolution." *Physical Biology* 5 (1): 015008. https://doi.org/10.1088/1478-3975/5/1/015008.

2009. "Dynamical Patterning Modules: A 'Pattern Language' for Development and Evolution of Multicellular Form." *International Journal of Developmental Biology* 53 (5–6): 693–705. https://doi.org/10.1387/ijdb.072481sn.

2011. "Lamarck's Dangerous Idea." In *Transformations of Lamarckism: From Subtle Fluids to Molecular Biology*, edited by Snait B. Gissis and Eva Jablonka, 157–70. Cambridge: The MIT Press.

Niklas, Karl J., and Stuart A. Newman. 2013. "The Origins of Multicellular Organisms." *Evolution & Development* 15 (1): 41–52. https://doi.org/10.1111/ede.12013.

Novick, Aaron. 2018. "The Fine Structure of 'Homology.'" *Biology & Philosophy* 33 (1–2): 6. https://doi.org/10.1007/s10539-018-9617-3.

2019. "Cuvierian Functionalism." *Philosophy, Theory, and Practice in Biology* 11: 20190821. https://doi.org/10.3998/ptpbio.16039257.0011.005.

Novick, Aaron, and W. Ford Doolittle. 2019. "How Microbes 'Jeopardize' the Modern Synthesis." *PLOS Genetics* 15 (5): e1008166. https://doi.org/10.1371/journal.pgen.1008166.

Nunes, Maria D. S., Saad Arif, Christian Schlötterer, and Alistair P. McGregor. 2013. "A Perspective on Micro-Evo-Devo: Progress and Potential." *Genetics* 195 (3): 625–34. https://doi.org/10.1534/genetics.113.156463.

Odling-Smee, F. John, Kevin N. Laland, and Marcus W. Feldman. 1996. "Niche Construction." *The American Naturalist* 147 (4): 641–48. https://doi.org/10.1086/285870.

Ohta, Tomoko. 1992. "The Nearly Neutral Theory of Molecular Evolution." *Annual Review of Ecology and Systematics* 23: 263–86.

Okasha, Samir. 2006. *Evolution and the Levels of Selection*. Oxford: Oxford University Press.

Ospovat, Dov. 1981. *The Development of Darwin's Theory: Natural History, Natural Theology, and Natural Selection, 1838–1859*. Cambridge: Cambridge University Press.

Outram, Dorinda. 1986. "Uncertain Legislator: Georges Cuvier's Laws of Nature in Their Intellectual Context." *Journal of the History of Biology* 19 (3): 323–68. https://doi.org/10.1007/BF00138285.

Owen, Richard. 1848. *On the Archetype and Homologies of the Vertebrate Skeleton*. London: Richard and John R. Taylor.

2007 (1849). *On the Nature of Limbs: A Discourse*. Edited by Ron Amundson. Chicago: University of Chicago Press. https://press.uchicago.edu/ucp/books/book/chicago/O/bo5550070.html.

Paley, William. 2009 (1803). *Natural Theology: Or, Evidences of the Existence and Attributes of the Deity, Collected from the Appearances of Nature*. 6th ed. Cambridge: Cambridge University Press.

Peter, Isabelle S., and Eric H. Davidson. 2011. "Evolution of Gene Regulatory Networks Controlling Body Plan Development." *Cell* 144 (6): 970–85. https://doi.org/10.1016/j.cell.2011.02.017.

2015. *Genomic Control Process: Development and Evolution*. San Diego: Academic Press. https://resolver.caltech.edu/CaltechAUTHORS:20151001-085324137.

Pigliucci, Massimo. 2007. "Do We Need an Extended Evolutionary Synthesis." *Evolution* 61 (12): 2743–49.

2019. "Causality and the Role of Philosophy of Science." In *Evolutionary Causation: Biological and Philosophical Reflections*, edited by Tobias Uller and Kevin N. Laland, 13–28. Cambridge, MA: The MIT Press.

Prince, Victoria E., and F. Bryan Pickett. 2002. "Splitting Pairs: The Diverging Fates of Duplicated Genes." *Nature Reviews Genetics* 3 (11): 827–37. https://doi.org/10.1038/nrg928.

Provine, William B. 2001 (1971). *The Origins of Theoretical Population Genetics*. Chicago: University of Chicago Press.

Raff, Rudolf A. 1996. *The Shape of Life*. Chicago: University of Chicago Press. https://press.uchicago.edu/ucp/books/book/chicago/S/bo3683614.html.

Reeve, Hudson Kern, and Paul W. Sherman. 1993. "Adaptation and the Goals of Evolutionary Research." *The Quarterly Review of Biology* 68 (1): 1–32. https://doi.org/10.1086/417909.

Reis, Claudio Ricardo Martins dos, and Leonardo Augusto Luvison Araújo. 2020. "Extended Evolutionary Synthesis: Neither Synthesis Nor Extension." *Biological Theory* 15 (2): 57–60. https://doi.org/10.1007/s13752-020-00347-6.

Rudwick, Martin J. S. 1997. *Georges Cuvier, Fossil Bones, and Geological Catastrophes*. Chicago: University of Chicago Press. www.press.uchicago .edu/ucp/books/book/chicago/G/bo3631234.html.

Rupke, Nicolaas. 2009. *Richard Owen: Biology without Darwin*. Revised ed. Chicago: University of Chicago Press.

Russell, Edward Stuart. 1982 (1912). *Form and Function: A Contribution to the History of Animal Morphology*. Chicago: University of Chicago Press.

Salazar-Ciudad, Isaac. 2021. "Why Call It Developmental Bias When It is Just Development?" *Biology Direct* 16 (1): 3, 1–13. https://doi.org/10.1186/ s13062-020-00289-w.

Sansom, Roger. 2008. "The Nature of Developmental Constraints and the Difference-Maker Argument for Externalism." *Biology & Philosophy* 24 (4): 441–59. https://doi.org/10.1007/s10539-008-9121-2.

Sapp, Jan. 2009. *The New Foundations of Evolution: On the Tree of Life*. Oxford: Oxford University Press.

Schank, Jeffrey C., and William C. Wimsatt. 1986. "Generative Entrenchment and Evolution." *PSA: Proceedings of the Biennial Meeting of the Philosophy of Science Association* 1986 (2): 33–60. https://doi.org/10 .1086/psaprocbienmeetp.1986.2.192789.

Schindewolf, Otto H. 1993 (1950). *Basic Questions in Paleontology: Geologic Time, Organic Evolution, and Biological Systematics*. Edited by Wolf-Ernst Reif. Translated by Judith Schaefer. Chicago: University of Chicago Press.

Simpson, George Gaylord. 1984 (1944). *Tempo and Mode in Evolution*. Columbia Classics ed. New York: Columbia University Press.

Sober, Elliott. 1980. "Evolution, Population Thinking, and Essentialism." *Philosophy of Science* 47 (3): 350–83. https://doi.org/10.1086/288942.

Soshnikova, Natalia, Romain Dewaele, Philippe Janvier, Robb Krumlauf, and Denis Duboule. 2013. "Duplications of Hox Gene Clusters and the Emergence of Vertebrates." *Developmental Biology* 378 (2): 194–99. https://doi.org/10.1016/j.ydbio.2013.03.004.

Speijer, Dave. 2011. "Does Constructive Neutral Evolution Play an Important Role in the Origin of Cellular Complexity?" *BioEssays* 33 (5): 344–49. https://doi.org/10.1002/bies.201100010.

Stellwag, Edmund J. 1999. "Hox Gene Duplication in Fish." *Seminars in Cell & Developmental Biology* 10 (5): 531–40. https://doi.org/10.1006/scdb .1999.0334.

Sterelny, Kim. 2000. "Development, Evolution, and Adaptation." *Philosophy of Science* 67 (September): S369–87. https://doi.org/10.1086/392832.

Stern, David L. 2000. "Perspective: Evolutionary Developmental Biology and the Problem of Variation." *Evolution* 54 (4): 1079–91. https://doi.org/10.1111/j.0014-3820.2000.tb00544.x.

Stern, David L., and Virginie Orgogozo. 2008. "The Loci of Evolution: How Predictable is Genetic Evolution?" *Evolution* 62 (9): 2155–77. https://doi.org/10.1111/j.1558-5646.2008.00450.x.

Stewart, Thomas A., Ramray Bhat, and Stuart A. Newman. 2017. "The Evolutionary Origin of Digit Patterning." *EvoDevo* 8 (1): 21 1–7. https://doi.org/10.1186/s13227-017-0084-8.

Stoltzfus, Arlin. 1999. "On the Possibility of Constructive Neutral Evolution." *Journal of Molecular Evolution* 49 (2): 169–81. https://doi.org/10.1007/PL00006540.

2012. "Constructive Neutral Evolution: Exploring Evolutionary Theory's Curious Disconnect." *Biology Direct* 7 (1): 35, 1–13. https://doi.org/10.1186/1745-6150-7-35.

2021. *Mutation, Randomness, and Evolution*. Oxford: Oxford University Press.

manuscript. "Grounding Internalism in the Population Genetics of the Introduction Process." www.molevol.org/grounding-internalism/.

Stoltzfus, Arlin, and Kele Cable. 2014. "Mendelian-Mutationism: The Forgotten Evolutionary Synthesis." *Journal of the History of Biology* 47 (4): 501–46. https://doi.org/10.1007/s10739-014-9383-2.

Stopper, Geffrey F., and Günter P. Wagner. 2007. "Inhibition of Sonic Hedgehog Signaling Leads to Posterior Digit Loss in Ambystoma Mexicanum: Parallels to Natural Digit Reduction in Urodeles." *Developmental Dynamics* 236 (1): 321–31. https://doi.org/10.1002/dvdy.21025.

Sultan, Sonia E. 2015. *Organism and Environment: Ecological Development, Niche Construction, and Adaptation*. Oxford: Oxford University Press.

Svensson, Erik. 2021. "The Structure of Evolutionary Theory: Beyond Neo-Darwinism, Neo-Lamarckism and Biased Historical Narratives about the Modern Synthesis." *EcoEvoRxiv*. https://doi.org/10.32942/osf.io/gjf8s.

Thompson, D'Arcy Wentworth. 1992 (1942). *On Growth and Form*. New York: Dover.

Uller, Tobias, Nathalie Feiner, Reinder Radersma, Illiam S. C. Jackson, and Alfredo Rago. 2020. "Developmental Plasticity and Evolutionary Explanations." *Evolution & Development* 22 (1–2): 47–55. https://doi.org/10.1111/ede.12314.

Van Fraassen, Bas C. 1980. *The Scientific Image*. Oxford: Clarendon Press.

2002. *The Empirical Stance*. New Haven: Yale University Press.

Wagner, Andreas. 2011. *The Origins of Evolutionary Innovations: A Theory of Transformative Change in Living Systems*. Oxford: Oxford University Press.

Wagner, Günter P. 1989. "The Biological Homology Concept." *Annual Review of Ecology and Systematics* 20 (1): 51–69. https://doi.org/10.1146/annurev.es.20.110189.000411.

2007. "The Developmental Genetics of Homology." *Nature Reviews Genetics* 8 (6): 473–79. https://doi.org/10.1038/nrg2099.

2014. *Homology, Genes, and Evolutionary Innovation*. Princeton: Princeton University Press. https://press.princeton.edu/books/hardcover/9780691156460/homology-genes-and-evolutionary-innovation.

Wagner, Günter P., and Kurt Schwenk. 2000. "Evolutionarily Stable Configurations: Functional Integration and the Evolution of Phenotypic Stability." In *Evolutionary Biology*, edited by Max K. Hecht, Ross J. Macintyre, and Michael T. Clegg, 155–217. Boston: Springer US. https://doi.org/10.1007/978-1-4615-4185-1_4.

Wake, David B., Gerhard Roth, and Marvalee H. Wake. 1983. "On the Problem of Stasis in Organismal Evolution." *Journal of Theoretical Biology* 101 (2): 211–24. https://doi.org/10.1016/0022-5193(83)90335-1.

Webster, Gerry, and Brian C. Goodwin. 1982. "The Origin of Species: A Structuralist Approach." *Journal of Social and Biological Structures* 5 (1): 15–47. https://doi.org/10.1016/S0140-1750(82)91390-2.

1996. *Form and Transformation: Generative and Relational Principles in Biology*. Cambridge: Cambridge University Press.

West-Eberhard, Mary Jane. 2003. *Developmental Plasticity and Evolution*. Illustrated ed. Oxford: Oxford University Press.

Whyte, Lancelot L. 1960. "Developmental Selection of Mutations." *Science* 132 (3432): 954. https://doi.org/10.1126/science.132.3432.954.

Wideman, Jeremy G., Aaron Novick, Sergio A. Muñoz-Gómez, and W. Ford Doolittle. 2019. "Neutral Evolution of Cellular Phenotypes." *Current Opinion in Genetics & Development* 58–59 (October): 87–94. https://doi.org/10.1016/j.gde.2019.09.004.

Wiens, John J., and Jason T. Hoverman. 2008. "Digit Reduction, Body Size, and Paedomorphosis in Salamanders." *Evolution & Development* 10 (4): 449–63. https://doi.org/10.1111/j.1525-142X.2008.00256.x.

Wimsatt, William C. 2013. "Evolution and the Stability of Functional Architectures." In *Functions: Selection and Mechanisms*, edited by Philippe Huneman, 19–41. Synthese Library. Dordrecht: Springer Netherlands. https://doi.org/10.1007/978-94-007-5304-4_2.

Winther, Rasmus Grønfeldt. 2015. "Evo-Devo as a Trading Zone." In *Conceptual Change in Biology: Scientific and Philosophical Perspectives on Evolution and Development*, edited by Alan C. Love, 459–82. Dordrecht: Springer Netherlands. https://doi.org/10.1007/978-94-017-9412-1_21.

Witteveen, Joeri. 2015. "'A Temporary Oversimplification': Mayr, Simpson, Dobzhansky, and the Origins of the Typology/Population Dichotomy (Part 1 of 2)." *Studies in History and Philosophy of Science Part C: Studies in History and Philosophy of Biological and Biomedical Sciences* 54 (December): 20–33. https://doi.org/10.1016/j.shpsc.2015.09.007.

2016. "'A Temporary Oversimplification': Mayr, Simpson, Dobzhansky, and the Origins of the Typology/Population Dichotomy (Part 2 of 2)." *Studies in History and Philosophy of Science Part C: Studies in History and Philosophy of Biological and Biomedical Sciences* 57 (June): 96–105. https://doi.org/10.1016/j.shpsc.2015.09.006.

2018. "Typological Thinking: Then and Now." *Journal of Experimental Zoology Part B: Molecular and Developmental Evolution* 330 (3): 123–31. https://doi.org/10.1002/jez.b.22796.

Wray, Gregory A. 2007. "The Evolutionary Significance of Cis-Regulatory Mutations." *Nature Reviews Genetics* 8 (3): 206–16. https://doi.org/10.1038/nrg2063.

Yampolsky, Lev Y., and Arlin Stoltzfus. 2001. "Bias in the Introduction of Variation as an Orienting Factor in Evolution." *Evolution & Development* 3 (2): 73–83. https://doi.org/10.1046/j.1525-142x.2001.003002073.x.

Ziporyn, Brook. 2020. *Zhuangzi: The Complete Writings*. Indianapolis: Hackett.

Acknowledgments

The words are my own; many hands have shaped the thought. My gratitude first and foremost to James Lennox, who taught me to be a philosopher of science by teaching me to be a historian of science. For comments on various incarnations of this project, I thank Stuart Newman, Yasmin Haddad, Ford Doolittle and members of his research group, members of the University of Washington Philosophy of Science Reading Group, Mel Andrews, and an anonymous reviewer for Cambridge. I am also grateful for many discussions with Alan Love, Bill Wimsatt, and Günter Wagner. An early version of these ideas appeared as a chapter of my dissertation; many thanks to my committee: James Lennox, Mark Wilson, Sandra Mitchell, Mark Rebeiz, and James Woodward. Finally, much of this Element was written during summer retreats at the Holly House; I am grateful to Katie Creel for her benevolent peer pressure and to her mother, Kathleen Creel, for her hospitality. My thanks to you all, and to all those who, unnamed here, have shaped my intellectual development.

Dedicated to Peter Brötzmann and Keiji Haino, whose
The intellect given birth to here (eternity) is too young,
released 2022.09.12, carried this work to the summit

Cambridge Elements

Philosophy of Biology

Grant Ramsey
KU Leuven

Grant Ramsey is a BOFZAP research professor at the Institute of Philosophy, KU Leuven, Belgium. His work centers on philosophical problems at the foundation of evolutionary biology. He has been awarded the Popper Prize twice for his work in this area. He also publishes in the philosophy of animal behavior, human nature, and the moral emotions. He runs the Ramsey Lab (theramseylab.org), a highly collaborative research group focused on issues in the philosophy of the life sciences.

Michael Ruse
Florida State University

Michael Ruse is the Lucyle T. Werkmeister Professor of Philosophy and the Director of the Program in the History and Philosophy of Science at Florida State University. He is Professor Emeritus at the University of Guelph, in Ontario, Canada. He is a former Guggenheim fellow and Gifford lecturer. He is the author or editor of over sixty books, most recently *Darwinism as Religion: What Literature Tells Us about Evolution; On Purpose; The Problem of War: Darwinism, Christianity, and Their Battle to Understand Human Conflict*; and *A Meaning to Life.*

About the Series

This Cambridge Elements series provides concise and structured introductions to all of the central topics in the philosophy of biology. Contributors to the series are cutting-edge researchers who offer balanced, comprehensive coverage of multiple perspectives, while also developing new ideas and arguments from a unique viewpoint.

Cambridge Elements

Philosophy of Biology

Elements in the Series

How to Study Animal Minds
Kristin Andrews

Model Organisms
Rachel A. Ankeny and Sabina Leonelli

Comparative Thinking in Biology
Adrian Currie

Social Darwinism
Jeffrey O'Connell and Michael Ruse

Adaptation
Elisabeth A. Lloyd

Stem Cells
Melinda Bonnie Fagan

The Metaphysics of Biology
John Dupré

Facts, Conventions, and the Levels of Selection
Pierrick Bourrat

The Causal Structure of Natural Selection
Charles H. Pence

Philosophy of Developmental Biology
Marcel Weber

Evolution, Morality and the Fabric of Society
R. Paul Thompson

Structure and Function
Rose Novick

A full series listing is available at: www.cambridge.org/EPBY

Printed in the United States
by Baker & Taylor Publisher Services

Printed in the United States
by Baker & Taylor Publisher Services